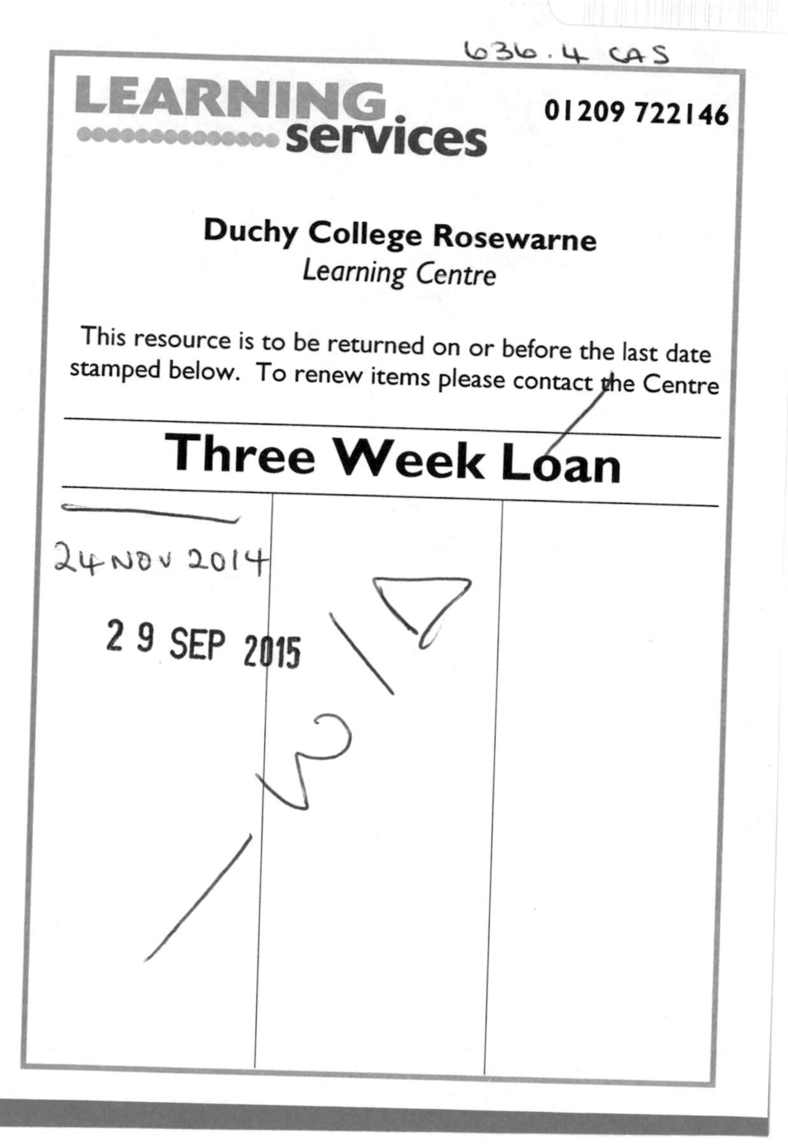

Broad Leys Publishing Limited

STARTING WITH PIGS

First edition: 2001. Reprinted 2004 and 2005.

Published by Broad Leys Publishing Ltd.

Printed by Biddles Ltd.

A catalogue record for this book is available from the British Library.

ISBN: 0 906137 29 2

Outside front cover photograph: Katie Thear. Outside back cover: Andy Case

Broad Leys Publishing Ltd
1 Tenterfields, Newport, Saffron Walden, Essex CB11 3UW.
Tel/Fax: 01799 541065. E-mail: kdthear@btinternet.com
Website: www.blpbooks.co.uk

Contents

Foreward

The farmer's bookshelf contains many volumes on the keeping of livestock which, over the years, have become progressively more scientific, detailed and lengthy. Go back to the era between the wars and you will find farming books written with enormous enthusiasm, the personality of the author popping through in interesting anecdotes. These sort of books bring their subject to life, satisfying the normal relevant enquiries of the eager amateur. This is one such book.

Andy and Maureen Case have kept pigs - all sorts of pigs - for many years. They have bred and reared them indoors, out of doors and even in their own home! There are not many summer showgrounds in the South West that have not seen them being awarded rosettes.

The author has his book firmly planted on the ground and by continuous osmosis has absorbed, over the years, enough pig lore to pale any passing porker. This knowledge has been distilled into a very readable and well laid out trip through all that one would wish to know about the keeping of pigs.

As one with but a modest acquaintance with the porcine world, I rush in and thumb through it for guidance regularly and strongly recommend the reader to do the same; you will find it most helpful.

Lord Seaford, West Knoyle, Wiltshire.

Acknowledgements

Dr. J.R. Walters, B.Sc., M.Phil., Ph.D., C.Biol., M.I.Biol. for help with genetics.
David G Lacy B.Vet. Med. M.R.C.V.S. for his help with veterinary advice.
All the 'Old Men' from whom I absorbed practical pig knowledge over the years.
To my wife, Maureen, who kept nagging me to get this book finished and who has been, and still is, a constant help to me and our pigs.

Some of the author's Kunekunes

Tamworth sow with her litter of piglets in an outdoor environment. The fencing is made up of posts with well-dug in pig netting to avoid 'burrowing under'. *Photo: Author*

Preface

This little book was born out of necessity. I found that people would ring me regarding the purchase of pigs. I would then spend over an hour advising them on the keeping of pigs, they would ask my prices and that would be the last I would hear of them! Two or three months later I would discover that they had bought their pigs from somebody else. It was a bit galling, to say the least.

At my wife's suggestion I decided that it would be a good idea to write down my knowledge of pig farming as a short book. I realised that it was essential to use straightforward language and to write in a friendly way for the first time pig keeper. Most books on the subject are written for the large scale farmer. They are intensive and scientific, and there is too much detail for the beginner.

This book is written by a small scale pig keeper for those who are interested in becoming proficient in keeping small herds. The headings and short chapters are arranged so that the novice can refer to essential topics with ease.The only real way to learn, of course, is to do it. 'Pig sense' cannot be acquired solely by reading. (This is immediately apparent when moving young pigs from one paddock to another! Please do not be put off, however. Obtain your pigs and enjoy them!)

This book then, is based on my practical experiences in the management of rare breeds of British pigs, as well as with the New Zealand Kunekune breed.

Andy Case, 2001.

Pigs in history

The Wild Boar, *Sus scrofa*, the ancestor of the domesticated pig. Harris, 1881.

The 'Old English' hog is a bit of a misnomer as it was essentially the same domestic pig as that kept by the Celtic races of Europe. It showed little development from its ancestral Wild Boar and took a long time to fatten. Harris. 1881.

The importation of Chinese pigs in the 17th century meant that pigs could be fattened more quickly, and the meat was more succulent.

Pigs have been kept by most of the great civilizations of the past. Here, a coin of Abacaenum from Sicily shows a boar and acorn, depicting the forest of oaks that offered pasture to the herds of swine.

The swineherd was responsible for taking pigs to pannage (to glean acorns and beechmasts in the forests).

The 'cottager's pig' was a common sight until the 1960s. After virtually disappearing as a practice, it is now more common again to keep a few household pigs.

(Old illustrations computer enhanced by Broad Leys Publishing Ltd)

Introduction

"They are great softeners of the temper, and promoters of domestic harmony".
(William Cobbett on Keeping Pigs - *Cottage Economy*, 1821).

Pigs have been kept by man for thousands of years and are found all over the world. Remains of a small, Eastern type of pig have been found at sites dating back to early Neolithic times. During the later Neolithic period however, they appear to have been replaced by animals of a more European or Eurasian type - ancestors of the Wild Boar, *Sus scrofa*.
The Wild Boar was widely hunted all over Europe, and is depicted in paintings, statuary, mosaics and coins, as well as being referred to in myths, legends and literature. The domestic pig is a descendant of the Wild Boar, as William Youatt pointed out in his book, *The Pig*, in 1847: *"No-one can for a moment doubt that it is the parent stock from which the domestic breeds of swine originally spring."*
The 'Old English hog' that Harris referred to in 1881 was similar in type to that kept throughout medieval times and descended from the pig kept by the Celtic races of Europe. It showed little development from its ancestral Wild Boar, apart from its curly tail, and took a long time to fatten.

Finding food for livestock was a perennial problem, particularly in winter when it was difficult enough for the peasants to feed their families, let alone their animals. For this reason, most pigs, apart from breeding stock, were slaughtered and salted for the winter. Every village had its swineherd in the past. He (or she) was responsible for taking pigs to pannage. This was the practice of allowing them to glean acorns and beechmasts in the forests, an important source of food for them. In the *Domesday Book*, the value of woodland was estimated according to the number of hogs that it would support.

To improve the speed of maturity, succulence and smaller size of our indigenous breeds, several colours of Chinese pigs were imported in the 17th century. These crosses eventually produced breeds such as the Berkshire and Middle White. The Landrace introduced a longer, leaner pig. Nowadays modern hybrids bred for large scale farming bear little resemblance to the pig kept and fattened a century or so ago, but an increasing number of small farmers are keeping the old breeds and hybrids - in a more traditional way.

With the mix of native breeds through the centuries of agricultural improvement, the pig has a complex genetic background. But we have not lost the special place in our hearts that we hold for the pig. He is anatomically the same as us, and more intelligent than other domestic farm animals, which explains why we find such pleasure in keeping pigs and why so many hanker to do so.

An old method of housing pigs was to utilise a lean-to from the human dwelling as illustrated by this Tudor house reconstruction in Sussex. The ladder allowed chickens to roost above. *Photo: Katie Thear*

A more satisfactory (and hygienic) method was to separate the pigs from the human dwelling. This is an old fashioned pigsty and run made of local stone in Wales. *Photo: Katie Thear*

Type of enterprise

"This little piggy went to market, this little piggy stayed at home".
(Traditional Nursery Rhyme)

It is assumed that readers of this book are potential *small-scale* pigkeepers, interested in keeping a few pigs in humane, non-intensive conditions. For most people, the choice of breeds will tend to be the more traditional ones, not only because there is more possibility of supplying the quality end of the market, through Farmers' Markets, Rare Breeds Survival Trust initiatives or Organic schemes, but also because it is helping to conserve endangered breeds.

Before starting, however, it is important to have knowledge of the various sectors of pigkeeping so that adequate planning is possible. There are essentially five areas of activity that are appropriate to the small-scale pigkeeper:

Pedigree breeding

This involves breeding and selling potential breeding stock. The choice will be one of the recognised pure breeds and the first essential is to obtain top quality, registered breeders. (Registered animals are those that are registered with the appropriate breed society as being pure-bred and from named parents).

A considerable investment is required to buy high quality animals. A cheaper alternative is to buy young, potential breeders and selectively breed from them in order to improve the line. This will obviously take some time but it is a good idea to show good examples of the breed at agricultural shows. In this way, a good reputation can be built up.

Breeding to sell weaners

This is where pigs are bred so that newly weaned pigs are sold. The parents may or may not be pure bred. A pure breed, for example, is frequently crossed in order to produce commercial crosses that tend to grow more quickly.

Breeding and fattening

This is where the weaners are kept and raised until they are ready for slaughter. This may be at pork weight (porkers) or for older pigs, at bacon stage (baconers).

The scale of activity here may be purely to supply the home freezer or to sell via Farmers' Markets.

Buying in store pigs and fattening

This is where piglets are bought in as weaners or slightly older and raised to slaughter. Again, this may be for the household or to sell. Great care needs to be exercised when buying in new stock, to ensure that it is disease-free.

Keeping pet pigs

A relatively recent development, this is where pigs are kept for interest and possibly for showing.

A pigkeeping activity may involve one, several, or all of the above, for they are not mutually exclusive. It is essential, however, to observe the regulations that apply to *anyone* keeping pigs, on any scale. This applies equally to domestic and commercial pigkeepers.

- Anyone keeping a pig must register with the local *Animal Health Office* (AHO).
- No pig can be moved without a Movement Licence or declaration of its status.

Further information and more specific details on regulations are available in the *Reference Section*.

Some terminology

As this is a book for beginners, it is appropriate to define some of the more common terminology that applies to pigs. A more complete round-up of definitions will be found in the *Glossary*.

Boar: An uncastrated male pig

Sow: A female pig that has had one weaned litter.

Gilt: A female pig up to the time her first litter is weaned.

Piglet: Young pig up to the weaning stage.

Weaner: A piglet that has ceased feeding from the mother.

Porker: A pig raised for pork.

Baconer: A pig kept for a longer period in order to produce bacon.

Store pig: A pig that is not being fattened when sold but may be raised for breeding or subsequent fattening.

Pure breed: A breed that when a male and female of that breed are crossed will always produce young like themselves.

Cross-breed: The young from a mating of two different breeds.

Hybrid: A commercial breed that has been developed from several breeds, using productive strains.

Housing

"In truth the pig, if given the right environment,
is by nature a clean and intelligent animal".
(Ranken Bushby - *The Complete Book of Raising Livestock and Poultry*, 1980)

Pigs can be kept on any type of land, in any part of the country, including good or poor quality land, hill or vale. Rough hill land is as good as any, being naturally free draining. Flat heavy clay soil is acceptable when the ground is dry, as long as there is provision to shut off the pigs from it in the winter or when it is too wet. (When the sun shines on a frosty winter morning, however, let them out for exercise; they will benefit greatly). Very little land is needed just to put up a pig sty, if the intention is only to fatten a few weaners for the freezer. If this is the case (and this is a good introduction) you need not have any land to run them outside. It is my belief and indeed, my practice to fatten (finish) weaners indoors in winter when it is cold. Weaners running outside will use all their food just to keep warm and will take ages to come to pork weight, and at too great a cost. In summertime when it is too hot, the same weaners will not eat enough because they do not feel like it. Far better for those being raised for meat to be inside, as long as there is an outside exercise yard.

To keep weaners some preparation is necessary. Little piglets will grow into large and very strong porkers with appetites to match.You notice that I refer to pigs in the plural. A single pig is a sad animal as pigs are gregarious and do so much better in pairs or groups.

Welfare

It is recommended that all prospective pigkeepers read the booklet *Codes of Recommendations for the Welfare of Livestock: Pigs*. (See *Reference* section). The salient points are as follows:

- Comfort and shelter.
- Readily accessible fresh water and a diet to maintain animals in full health and vigour.
- Freedom of movement.
- The company of other animals, particularly of like kind.
- The opportunity to exercise most normal patterns of behaviour.
- Light during the hours of daylight, and lighting readily available to enable the animals to be inspected at anytime.
- Flooring which neither harms the animals, nor causes undue strain.
- The prevention, or rapid diagnosis and treatment, of lice, injury, parasite infestation and disease.
- The avoidance of unnecessary mutilation.
- Emergency arrangements to cover outbreaks of fire, the breakdown of essential mechanical services and the disruption of supplies.

Lying space per pig by live weight

These figures do not include space for dunging and exercise.

Live Weight (Kilograms)	*Area* (Square metre)
20	0.15
40	0.25
60	0.35
80	0.45
100	0.50

Housing requirements

A pig house needs to be draught-free but well ventilated otherwise respiratory problems may ensue. It requires an adequate and uniform temperature which will normally be the case if the house is well insulated. The pigs themselves will then provide the warmth.

Most heat escapes through the roof. If a high-roofed outbuilding is being adapted for pigs, a false ceiling can be inserted to cut down on heat loss.

An insulated floor also makes a big difference. In an outbuilding or sty with run, this may be a layer of insulation under concrete. In fact, if a new concrete floor is to be laid, polystyrene tiles or even egg cartons are very effective. Equally important is that the flooring should not be too smooth otherwise the pigs' feet do not get enough grip and they can fall and hurt themselves. 'Combing' the wet concrete into slight ridges is an effective way of achieving a satisfactory surface.

The provision of straw on the floor enables the pigs to arrange it themselves to make a warm, comfortable bed and to prevent injury to their hocks.

Sties or adapted outbuildings need outside exercise yards. Here, good drainage is essential so the run needs to slope slightly away from the house. The door from the run into the house needs to be strong, with bolts on the outside unless, of course, the pigs are in outdoor, free-range houses.

These are general requirements for all pig housing. Adaptations for farrowing and so on, are indicated in the appropriate chapters.

Home-made housing

The simplest form of house is made from three 180 cm (6 ft) sheets of galvanised iron nailed to four posts hammered into the ground. Four 240 cm (8 ft) sheets for the roof makes a perfectly good house for older and adult pigs. In summer some straw bales on the roof will keep it cool, and conversely, the same layer of bales will keep the pigs warm in the winter. In adverse conditions tack a piece of old carpet over the front to keep your pigs really snug.

Wooden pallets stuffed with straw and wired together make a good temporary home for pigs in the summer. The roof can be chicken netting stretched and stapled across the top and heaped with straw. If one starts the heap in the middle of the

This is a plastic ark on two layers of straw bales to give store pigs shade in hot weather.
Photos: The author

Details of construction of a renovated ark

An excellent method of cooling a metal farrowing ark with straw bales.
Note the fender to keep tiny piglets from straying.

wire netting and keeps adding to it so that it spills outwards from the top, it will shed the rain like a thatched roof, provided you tie on an old piece of chicken netting to stop it blowing away.

If you are offered an old pig ark, do go and see it. Provided the curved galvanised sheets are not rusted through in holes, and even if the edges at ground level are a bit rusty, have it! It is a simple matter to rebuild it with 1.25 cm (0.5 in) ply wood ends and new timber. (1 inch = approx 2.5cm). Here's how to do it:

Cut 7.5 cm x 5 cm (3 in x 2 in) sawn timber for the frame, to fit the length and width of the curved sheets when bolted together. Drill holes in the ends and with 11 cm (4.5 in) long 1 cm ($^3/_8$ in) coach bolts, fix the long timbers on top of the shorter end timbers.

Now stand your 240 cm x 120 cm (8ft x 4ft) sheet of 1.25 cm (0.5in) exterior plywood at one end of the curved roof. Mark the curve of the roof onto the sheet of plywood with a pencil and cut out the end wall with a jig-saw.

For the front end use the end wall shape as a template but cut it in half so that you will have one side only covered in the front and the other side left open for entry. Now nail the end wall to the frame along its bottom edge, and also the front half to the front timber along its bottom edge.

Nail a piece of 80 x 5 cm (32 in x 2 in) wood vertically in the middle of the back wall finishing 5cm (2 in) below the centre top edge. Do the same at the front.

Now saw a length of 7.5 x 5 cm (3 in x 2 in) to fit between the ends on top of the two vertical timbers so that it is at roof level, and secure with 10cm (4in) nails.

Nail triangles of plywood to brace the upright timber with the roof timber as shown on page 13.

Roll the whole thing over on its roof and nail the end frames to the bottom ends of the vertical timbers, again with 10 cm (4in) nails.

The whole thing is greatly strengthened by screwing shelf brackets to the long side timbers and drilling holes in the plywood ends to bolt the other arm of the bracket to them. All one needs to do now is to fix the curved sheets to the frame. Use spring head nails, nailed in the 'valleys' of the corrugations at the bottom edges and also to the roof timber. It helps to drill if the nails are taken out first. One does not need to buy

Here a shed has been adapted as pig housing, with concrete runs for exercise. Each run is separated by a fence made of wooden pallets reinforced with galvanised iron sheets. *Photo: Katie Thear*

Plastic, moveable arks for the outdoor pig. *Photo: Carbery Plastics Ltd*

Wooden and galvanised iron field arks are used here for the free-ranging sows and piglets. *Photo: Author*

A modern pig house made of concrete blocks, with an outside exercise yard. The breeds shown here are (left to right) Tamworth, Berkshire and Gloucestershire Old Spots. *Photo: David Thear.*

curved galvanised sheets for the roof. Conventional flat sheets nailed lengthways work very well on the same framework, provided one nails 5 x 5 cm (2 in x 2 in) blocks 15cm (6 in) long all around the curved edge of the two ends to fix the sheets.

Start nailing on the bottom sheets on both sides and work up the sides to cap the top with the centre sheet and there you have an inexpensive pig ark.

An even simpler ark is made with six feet galvanised iron sheets held upright and nailed together as a 'ridge tent'. (See page 20). The triangular ends can be wood or galvanised iron sheets cut to fit. Leave a 60 x 90cm (2ft x 3ft) high doorway one end and a 60 x 60cm (2ft x 2ft) window at the other end that can be closed for winter use and opened for summer. If you fix three rails that run the length of the ark, the height of the bottom edge of the window, you can store two bales of straw there which would also keep the ark much warmer.

Pig arks

There are ready made pig arks available in a variety of materials, including corrugated iron, wood and plastic. Arks normally have curved roofs, as for example the ones made of curved iron sheets with wooden ends. Some arks are insulated and there is a considerable variation in price. Plastic ones are lighter and easily moved by turning them on their backs and sliding them across the ground as a sledge.

Arks with floors may harbour lice and provide a habitat for rats underneath them in winter time. It is important to move the ark onto clean ground and clear up or burn the old straw litter. (This is especially important with farrowing arks). When your ark is not in use turn it on its back for the weather to clean it.

Sheds and stables

A shed can be used as housing, but it needs to be at least 180 sq. cm (6 sq. ft) with a similar sized outside run. The floor must be concreted as pigs can be great diggers. It is of great benefit to the pigs to insulate the concrete floor, inside the shed with a layer of concrete 5cm (2in) deep then a 2.5cm (1in) layer of polystyrene insulation and about 3.5cm (1.5in) of concrete on top. This considerably warms up the living area and guarantees that the pigs will dung outside. Pigs are the cleanest of animals.

If the only available building is a conventional stable this is quite adequate. For preference it should be 3 x 3 metres (10ft x 10ft). Straw down a part of it for them to lie on and wet or put some soiled litter down near the door. This is normally where your pigs will prefer to dung. When your weaners are newly bought and are only young, say six weeks or eight weeks old, one can give them a straw bale to get behind for warmth and perhaps suspend a sheet of ply wood over the lying area as a false roof to help warm things up.

Pig sty

If you build a concrete block pig sty, it is important to have it facing the right way. Pigs will do much better if the sty faces south. If it is north-facing, they will not get enough sun; east-facing it will be too cold and west-facing will be too wet.

When deciding on the size of the building allow 1.3 square metres (4.5 sq.ft) of lying space per pig, and 35cm (14in) of trough space.This will be sufficient for pigs up to bacon weight. Do not be tempted, when expanding your pig accommodation, to build another line of sties facing the first. I maintain that all sties should face south for the greater good of the pigs. So, add on to the same line or have another line also facing south.

You may well find it beneficial to have pig pens of different sizes: 1.8 x 1.8 metres (6ft x 6ft), 1.8 x 2.4 metres (6ft x 8ft) and 1.8 x 3.6 metres (6ft x 12ft). I find, also, that you can never have too many pens. It is a good idea to be able to rest a pen between batches of pigs. This can then be hosed down and disinfected, allowing plenty of time for the floor to dry out completely. This is particularly important if farrowing sows indoors. Never allow a heavily in-pig sow to lie on a wet concrete floor, especially in cold weather.

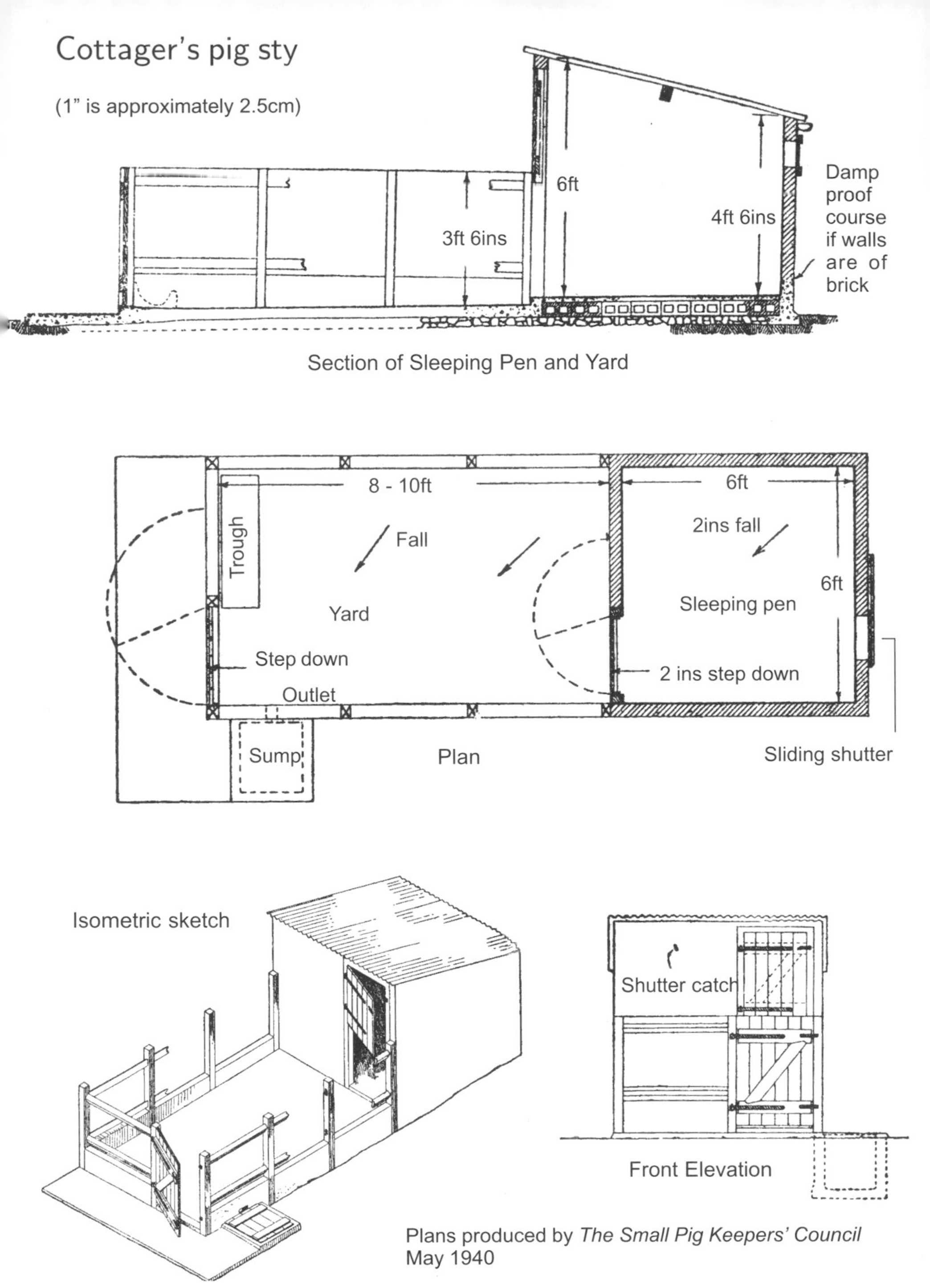

Plans produced by *The Small Pig Keepers' Council*
May 1940

Cottager's Pig Sty

The following plans were produced by *The Small Pig Keepers' Council* for use in wartime Britain.

Foundations*:* Portland cement concrete

Walls

1. Bricks in cement or lime mortar. (With brick walls a damp proof course should be provided).
2. Hollow building or breeze blocks, etc, in cement mortar, rendered externally.
3. Concrete cast *in situ* or precast blocks.
4. Timber framing covered externally with galvanised corrugated iron, weather boarding or vertical boarding, etc. Lining may be of boarding or corrugated iron.

Walls of Yard

1. Materials as described for walls of sleeping pen 1, 2 and 3.
2. Concrete or wood posts concreted in, with rails to receive boarding, slabs or off-cuts.
3. Galvanised corrugated iron and pig wire with concrete plinth.

Floors

Sleeping pen: Concrete on hard filling well rammed, hollow building blocks or land drains with 2 inches of concrete over.
Yard: Concrete 3 inches thick, laid to fall towards outlet to sump.

Roof

Boarding and bituminous felt, galvanised corrugated iron with suitable insulation, or thatch

Doors

1 inch T. &. G. boarding, ledged and braced, with heavy hinges bolted through ledges, and suitable barrel bolts.

Ventilator

Sliding shutter

Trough

Portable trough of galvanised iron or fixed trough of concrete or glazed fireclay.

Painting

All woodwork should be treated with a preservative. Corrugated iron and other ironwork should be painted with bituminous or other suitable paint.

A temporary straw bale house with a wooden pallet and tarpaulin roof. Metal hurdles have been used as a boundary fence. *Photo: Katie Thear*

A small house of galvanised iron sheets arranged in a ring. It has a conical, thatched straw roof which is warm in winter and cool in summer. The pigs are Kunekunes. *Photo: Author*

A wooden pig kennel which can be moved to a different site when necessary. Flaps across the door allow passage while excluding draughts. The roof slopes backwards to shed rain away from the entrance. *Photo: Sturdy Stys Ltd.*

Fencing and pasture

"Few sights are prettier than a set of hogs, slick and shiny, on a rich, green pasture".
(Kelly Klober - *A Guide to Raising Pigs* - 1997)

The perimeter of the pig paddocks or boundary of the property should be fenced so that pigs cannot roam. This should be done properly with pig netting supported every three paces by wooden stakes. Half round, 2.5 x 10cm (1in x 4in) diameter tanalised stakes are better and cheaper than the round ones. Two to three inch round ones are not strong enough for big pigs.

The wire should be strained up very tightly.This is best done by first stapling up one end of the roll to fix it, and then at the other end, trap the netting between two pieces of 105cm (3ft 6in) long, 10 x 5cm (4in x 2in) timber bolted together by three coach bolts. Now stand the netting up, attach a rope to the top and bottom of your two pieces of timber, tie the rope to your tractor (or other vehicle) and very slowly pull up tight. Corner posts can be double braced for extra strength.

Start stapling up the wire to the posts that are out of the straight line first, finishing up by your straining vehicle. Let the strain off and then run out a single strand of barbed wire at ground level. Make a loop to attach a rope to it and pull it tight. The barbed wire is to stop the pigs rooting underneath the fence.

It is a good idea to run a strand of barbed wire at the top of your fence, 5cm (2in) above the pig netting. A sow trying to get through to the boar, and he straining to drag her into his paddock, could well slacken or distort the fence.

I always wire the top wire of the pig netting to the top barbed wire line in the middle, between the spaced posts and in two places on the bottom wire of the pig netting, to the ground line of barbed wire. This holds the whole thing together and stops the fence bulging outwards.

Division fences

Division fences can be a single strand of electric fence wire or fencing tape. The pigs can see tape better and it is easier to move stubborn ones because they can see that it has been taken away. This may not always be enough and a good tip is to straw down a pathway through the fence line.

Never use stranded electric fencing wire. If it gets tangled around a pig's foot it may cut it off. (I saw this happen once). It's a waste of time and money using electric *sheep* netting. If the power fails or piglets try going through it and get caught, the racket they kick up alarms the mother who rushes through it, dragging it down and squealing as she wraps it around her neighbour's ark and piglets.

Granted a single strand will not keep little pigs in, but larger pigs provided they are first trained to it, will happily stay behind it. Remember the thicker the gauge

Sturdy fencing is essential. Here a Tamworth investigates the possibility of the grass being greener on the other side. The fence is made up of posts and rails with pig netting.
Photo: Katie Thear.

of fencing wire, the better the conductivity. Two strands are better than one for more permanent fences; one strand 23cm (9in) off the ground and one 23cm (9in) above that. Buy 1.3cm (half inch) iron rod cut to 76cm (2ft 6in) lengths. By using plastic insulators that are manufactured to fit on and tighten onto iron fencing stakes and then hitting them 15cm (6in) or so into the ground with a lump hammer, you have some inexpensive fencing posts.

A mains electric fencer unit is best for semi-permanent fence lines because it gives a better, more consistent performance. It is also cheaper to run. Mains chargers should be protected from the weather. Alkathene insulated pipes are useful for taking electric wires around corners. Portable, battery charged chargers are also available. Remember to check that the battery has not run down. The grass where an electric fence is situated needs to be kept mown otherwise electrical 'shorting' may occur.

Some pigs are great jumpers. I have seen rampant gilts leap a whole line of electric fences.

If keeping breeding stock, as opposed to fattening pigs, it is best for them to run outside on grass. A couple of gilts can be kept in quite a small area, as long as it is divided into two. The recommended number is six sows per acre (0.4 hectare). Of course, more per acre of little pigs like the New Zealand Kunekune can be kept on

Two strands of electrified wire are better than one, and are certainly required for young pigs.
Photo: Katie Thear

the same area. Having divided the paddock, one moves the pigs back and forth as they eat off the grass. In winter use one side as a 'sacrifice' paddock and reseed it in the spring.

Pigs are not able to sweat so it is important for them to be able to cool themselves in hot weather. Light coloured pigs are also susceptible to sunburn. The provision of shade is important. Providing a shallow 'wallow' of water where they can cool themselves is also very effective, not to mention popular!

Rotation of Pasture

Grass made available in rotation	Moveable electric fencing used to control access	In winter use one side as a 'sacrifice' paddock. Reseed in the spring. Rotation of pasture after 12-18 months use breaks the life cycle of parasites.	— Permanent perimeter fencing
When grass is eaten, move the pigs	Rake vacant areas to encourage new growth		Six sows per acre (0.4 hectare) (See also page 45)

A wallow where pigs can cool themselves is essential in hot weather.
Photo: Cotswold Pig Company.

Maureen Case's great grandfather, Mr H. Francis with his herd of Gloucester Old Spots at East Knoyle, Wiltshire. *Photo: A. & M. Case*

Breeds

"A natural breed is a contradiction in terms"
(R. Patrick Wright, 1912)

There are a number of breeds of pig to choose from, which may or may not, be suitable, depending on the size or temperament, and the ability to live outside. Personally, I would always go for a rare breed. Traditional, rare breeds need to be conserved and it is mainly the small pig keepers who are playing such a vital role in this respect.

Traditional breeds

Middle White

The Middle White has been cared for by the *National Pig Breeders' Association* (now the *British Pig Association*) since the association was founded in 1884. Herd Books have been published since that date.

Robert Bakewell was one of the improvers who, with Chinese Pig blood and diligent crossing, reduced the size and increased the flesh and speed of maturity of the large white pigs of the North of England. Between the two world wars the Middle White was the most popular pig in Britain. Thousands were sent to London every week for the London pork trade.

The Middle White is a very sweet natured, medium sized pig, but, in my experience, tends to sulk if it cannot get its own way. It is a quick maturing pork breed but cannot be taken to bacon weight as it gets too fat. It produces succulent pork very quickly. I frequently produce small porkers at only 16 weeks.

British Lop

The British Lop is the rarest pig breed. It is based on the long, white lop eared pig of Cornwall and is similar to the Welsh and Landrace pigs. It is also similar in conformation, size and shape to the Large Black which is also from the West country, although it is leaner than the Large Black.

It is large, lop-eared, friendly and very noisy. British Lops make good mothers, and are very milky. I find that this breed needs crossing with the Middle White to speed the finishing process up to pork weight. Pure-bred it makes an exellent bacon pig of exceptional flavour.

Tamworth

The Tamworth is an ancient British breed of red or ginger colour and with a distinctive long nose. The latter is inherited from wild boar. It is a large pig which is very slow to mature. It has no Chinese or Neapolitan blood in its veins.

As marketing requirements changed some traditional breeds, such as this Lincolnshire Curly Coated pig, became extinct in Britain. (The curly coated Mangalitsa is still found in Hungary and there are plans to introduce it into the UK). Photo: *Marshland Marion* photographed by Parsons for Gresham's *Livestock of the Farm* published at the turn of the last century.

In recent years, wild boar have been farmed, as well as being used to produce 'Iron Age' crosses. *Photo: Katie Thear*

Tamworths should have fine shoulders and well filled hams. They are very hardy, resistant to sunburn, make good mothers, are vocal and will run like a race horse. They will eventually make lean pork but take months to get to bacon weight. Tamworths need careful feeding in the later stages otherwise the baconers will be too fat. This is one reason why so many run them outside. Many are kept in woodlands, an ideal situation for a pig and close to its natural environment.

Berkshire

The Berkshire is a medium sized pork breed of very smart appearance, being black with pricked ears and four white socks, a white blaze on its face and the tip of its tail. The short, broad body and concave face are inherited from the Chinese pigs imported in the 17th, 18th and 19th centuries.

They make exellent porkers at an early age and like the Middle White are best killed to produce a 36 kg (801b) deadweight carcase, and a killing out percentage of 76-8%. With both the Berkshire and the Middle White one should take advantage of their quick maturing attributes for a faster turnover and quicker throughput. I find that Berkshires can be naughty pigs but are good natured, make very good pork but are not for bacon generally. However, they are sometimes crossed with the Large White in order to produce bacon pigs.When crossed with a white boar all the offspring will be white.

A popular cross, particularly in New Zealand and Australia, used to be with a Tamworth boar (the offspring being dual purpose and ginger with black spots)

Large Black

The Large Black is another old breed which was found right across the south of England. It is large, lop-eared and very easy to keep in, as its ears restrict its sight to some degree. The sows are exceptional mothers and always do their piglets well. They are slow to mature with not very fleshy hams, although recently breeders have largely overcome this. They are bacon pigs and the offspring do very well for bacon if crossed with a Large White boar. Crossing with a Middle White also produces good porkers.

The crossbred pigs were highly prized until recent times, being the so-called 'Blue-Pig'.There was a good market for blue weaners for the bacon trade. Some considered them to be the best bacon pig of all, with their large musculature and leanness of fat.

Gloucester Old Spots

The Gloucester Old Spots breed has been around for roughly 200 years. It is said to have evolved in the lower Severn Valley where it was kept in orchards, feeding on the windfall apples. It is a very hardy, big, lop-eared pig, coarse haired and coarse boned. In my experience some individuals can also be coarse tempered. The Gloucester is always stubborn. It has good hams and is supposedly dual-purpose,

British Saddleback. *Photo: Katie Thear*

Middle White. *Photo: National Pig Breeders' Association*

pork or bacon but takes a fair time to get there. Their heyday was during the 1920s and 30s, then their popularity collapsed. Recently they have enjoyed quite a revival although they are no longer as spotted as they used to be.

British Saddleback

The British Saddleback (before 1967 known as the Saddleback) is a striking breed being black with a band of white behind its shoulder. The first crosses were made on the Isle of Purbeck in Dorset from an amalgamation of the old Essex and Wessex Saddlebacks. It quickly became popular so that by 1825 pigs were exported to the U.S.A. to found the American Hampshire.

The sows make exceptional mothers. They are docile, good tempered and milk very freely. Of all the rare breeds they are arguably the best for open-air management. They can forage for themselves and can pick up most of their living from grass alone. The breed is so hardy that no expensive housing is necessary, improvised shelters being all that is required. Like the Large Black, they are easy to keep but one needs to watch out when the sows have newly born piglets. They are very protective. Again, as with the Large Black, they need crossing to produce good porkers and baconers.

Large White

The large White pig deserves its title of a universal breed. It has been exported all over the world and is indeed, still to be found all over the globe. It was developed around Leeds in Yorkshire in the middle of the 19th century from the large coarse, type that lived in Yorkshire and the Northern counties. The foundation was in 1851 when a Mr.Tuley from Keighley exhibited the first one at the Royal Show. Such were his sucesses that other breeders followed his example and then many strains of white pigs were found.These were developed into the Large White, Middle White and Small White. Published records can be traced back to 1884.

The breed is typically a bacon one, although porkers can be sold before they get to bacon weight. It has length and lightness of shoulder. Although rather high on the leg, it possesses very well filled hams. Now widely used as a crossing sire, the Large White is prolific and hardy.

Oxford Sandy and Black

The Oxford Sandy and Black's history is pretty murky but it has links with the Old Berkshire and Tamworth. There is no doubt that it is very old and was definitely the Cottager's pig. The trouble was that it it was not until 1985 that a breed society was formed. It had never had one, or a herd book, but that apart, the Oxford is a delightful pig. It is truly dual purpose. Whatever weight you take the weaners to, they never get too fat, in my experience. It is for this reason and the fact they are so docile and trouble free to keep, that I always recommend them to the beginner.

Berkshire at the Devon County Show, 2000. *Photo: Author*

They are very hardy, good mothers and of smart appearance, being lop-eared, ginger or sandy in colour with black blotches, white blaze, white socks and tassel.

Buy Oxford Sandy and Blacks of good constitution, deep in the body. There are a few about that look like greyhounds. Do please breed from registered stock and never from prick-eared sows or black and white, or all ginger ones.

The Oxford boar used on both the Large Black and British Lop makes for excellent cross weaners, and I maintain that there is nothing to beat the flavour of this 'Plum Pudding' pig; a natural forager.

Welsh

This large, white, lop-eared pig classified as a commercial breed was kept for generations and bred along pure lines on Welsh Valley farms but was unknown outside its native land. In 1907 Youatt described the Welsh as a razor-backed, coarse haired, slow-maturing kind, unprofitable but capable of great improvement. And improve they did. After the second world war the Welsh became one of the foremost whites in Britain, good for pork or bacon and possessing superb hams. Recently, alas, the breed's numbers have dropped alarmingly due to the big pig breeding companies breeding their hybrids without Welsh blood. They thrive on little bought in feed, are docile and are wonderful mothers.

New Zealand Kunekune

The Kunekune (pronounced Kuneykuney) comes from New Zealand and has had a long association with the Maori people. It has a wonderful temperament and loves human contact and company. This delightful, little, hairy pig comes in a variety of colours and sizes.The very best though, are only 51-56 cm (20-22 in) at the shoulder as fully grown adults. They are short legged, stocky pigs with surberb hams and great depth of meat.

They were probably brought from Southern Asia by the whalers who fished the New Zealand waters and traded them with the Maori, who prized them highly because they did not roam and because of the quality and quantity of their meat and fat. The Maori name *kunekune*, means fat and round, with their distinctive tassels or Piri Piri under their chins; they have never been wild. So, having always been domesticated makes them ideal for pets or small pork. They should not be confused with the Captain Cooker pigs which have long snouts, are bigger, leaner and are long legged. The Kunekune is a short nosed, short legged, short bodied, tubby pig that matures very quickly. They make very succulent small porkers at only five months old.

Vietnamese Potbellied

The Vietnamese from Asia was exported to the U.S.A. and Europe, originally for laboratories in the 1960s. During the 1980s it became a fashionable city pet. It is a small pig of very wrinkled appearance and will become grossly fat. They are very precocious, reaching puberty at only three to four months of age. They are usually ready to start breeding at six months. I have never kept any but am told they can be killed for pork at a young age.

Commercial breeds

There are a number of commercial pig breeds that the small pig keeper will come across in the local livestock market, including the Large White and the Welsh already mentioned.

Landrace

The Landrace is a lop-eared and long-backed white pig. Imported from Denmark it has been bred for leanness, to cater for consumer demand. It has been widely used to cross with British breeds and to develop new strains.

The Landrace cross with the Large White is a very good combination, but not as hardy as the American Duroc breed.

Duroc

The hardy, American-bred Duroc makes the ideal terminal sire on Large White and Landrace crosses, in my view. It has very good conformation and, being coloured, does not get sunburnt. It also has enough back fat to do well outside. It is a docile breed as well as being prolific. I have kept and finished Duroc cross Large Blacks with great success. The Duroc blood in the Large Black crosses certainly speeds

The author's wife, Maureen, with a Kunekune piglet at the Royal Show. *Photo: Katie Thear*

Large Black sow and piglets. *Photo: Katie Thear*

them up to pork weight, giving them much deeper and fuller hams than a pure-bred Large Black would have. In my experience, the Duroc crossed with Middle White is successful for the same reasons.

Hampshire

Bred in the USA, the Hampshire is like a prick-eared Saddleback to look at and indeed was developed from the white belted, black pig from the New Forest in Hampshire.

Chester White

Also from the USA is the Chester or Improved Chester White, an amalgam of white breeds and types; so much so that from around 1850 until 1950 when it was established as a breed, there were getting on for a dozen breeds introduced. It has now emerged as a lop eared pig not unlike the Landrace.

There are also hybrid strains of pigs available, although many of these have been developed for more intensive conditions. However, with the increasing emphasis on outdoor pig production, there are also strains appropriate to these conditions being bred. Artificial insemination also provides an opportunity to use top rate commercial boars on traditional breed gilts to produce commercial crosses.

Vietnamese Potbellied pigs. *Photo: Katie Thear*

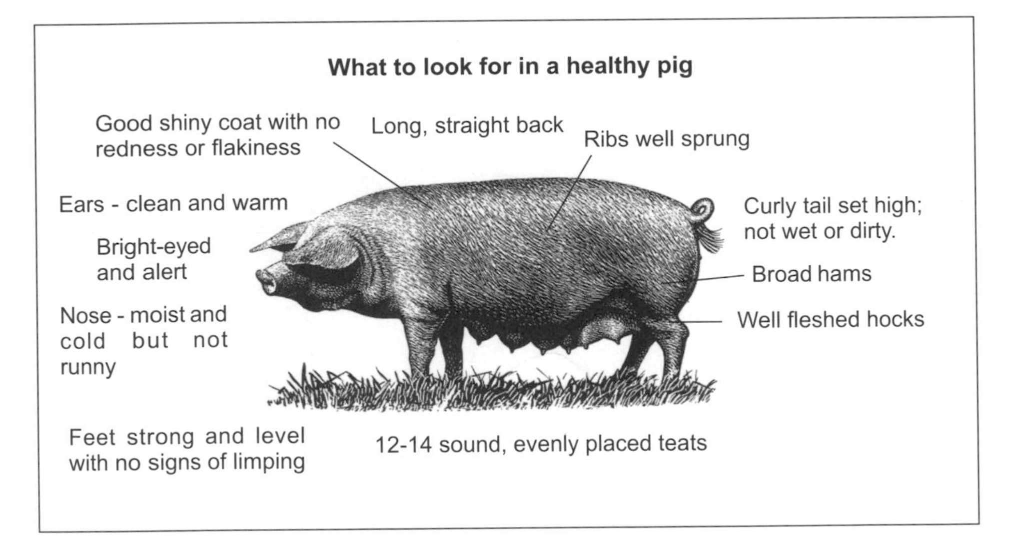

To tell the age of pigs

- A pig is over six months of age when it has its permanent corner incisors.
- It is over nine months when the tusks appear.
- It is over one year when the central and corner incisors are in wear.
- At 15 months the central teeth have pushed the 'nib' teeth out.
- At 18 months the central ones are in wear.
- After this, it's unreliable to try and tell the age of a pig by its teeth.

Buying stock

"Don't buy a pig in a poke!"
(Traditional saying)

One can buy eight-week weaners of coloured pigs very cheaply in the market. The reason why they are cheap is that the commercial pig farmer is not interested in them. However, the problem is that if buying in this way, you will have no idea of their health status. It is better to look in the local paper for advertisements under the *Animal* section. Telephone the vendor, explain what and how many you require, and go and see the pigs before you buy. In this way, you should end up with a much better and healthier sort of weaner, albeit for a little more money.

Buy registered or from registered stock. Why? Because it is simply a matter of economics. A pure bred, rare breed weaner with its papers will be more expensive. At least buy a rare breed pig that is 'birth notified' if you intend to breed from her. This is because you will not be able to register her or her offspring unless she has been notified in this way. When you sell weaners, registered ones are worth more than double those that are unregistered. You will have a chance of doubling your profit.

If you prefer, ring up and buy from a recognised breeder who advertises in smallholding and local or national farming publications. It is also good idea to get in touch with the *Rare Breeds Survival Trust (RBST)*. They will point you in the right direction.

It is important when buying pure bred stock to obtain the very best you can afford. This does not necessarily mean the most expensive price quoted. Look for a weaner that has a good shiny coat and a skin free of redness or flakiness. It should look alert, be bright in the eye and come to you. Do not buy any from a pen where there is very loose dung; so look out for wet or dirty tails. Above all, your prospective purchase should walk well. The 'clays' the horny foot, should be level with the pasterns strong and straight.

A tip worth knowing is to pick the weaners with a dimple just above their tails. They always finish quickest and best.

You can keep six adult pigs to the acre. When buying pedigree rare breed stock for your outdoor system be sure to buy from those that are kept outside or for most of the year to ensure they are hardy pigs. Never buy a Tamworth gilt that has a pink skin. Tamworths should have dark skins. Never buy a Berkshire gilt that has red skin; it should be black. In both instances the wrong coloured skins indicate that they definitely did not live out nor were they housed outside in arks.

Transporting your pigs home should be an easy operation. If you have bought little weaners, a stout box of ample proportions, bedded with newspaper and straw is adequate. Slip the piglets into it without taking it out of your hatchback or estate car. In warm weather leave the windows open in the back for plenty of ventilation. Remember too, that the sloping back window of a hatch-back on a sunny day acts like a greenhouse over the piglets and must be covered to stop the sun blazing through. Make sure that the box cannot move in transit.

Any pigs larger than weaners must be transported in a trailer. By law you must have a clean trailer that is cleaned out within 24 hours afterwards. If you cannot borrow one, then I'm afraid you'll have to hire one. There are companies that specialise in doing so. A two wheel trailer, preferably with a hard top is best. It should have a tailgate that lets down as a ramp.

Depending on the number and size of the pigs you intend to buy, you might have to come to a four wheel trailer. These usually have tail sidegates to facilitate loading. And believe me a four wheel trailer is much easier to manouvre and back than a two wheel one.If you can't back a trailer, it's a simple matter to unhitch your two wheel trailer, push it back by hand into the required position and hitch up again. It is far better in the hot summer months, when metal trailers can heat up like ovens, to move pigs very early in the morning or late evening when it's cooler. Always carry water and a drinking bowl for them.

Do not over-do the feeding before you transport pigs; perhaps do not feed at all until their journey's end. Little pigs in particular can, quite frequently be car-sick, particularly if they've been fed just before setting off. I will go into the art of loading pigs in a later chapter but it does require preparation and some thought. It is of utmost importance when arriving home with your pigs to put them in a house that is secure. Do not just let them out into the paddock for they might 'Woof-woof and away' down the other end through the wire and that will be the last you see of them, until that is, you are called from your bed by the police who have been alerted by someone who has picked them up in his headlights a mile and a half away. The time now is a quarter to two in the morning. Daylight sees them safely back in a new pigsty.

If you have bought young pigs do keep them in for a week until you get to know them and they get to know you and your voice. In any case, if you already have pigs on the place, it is sound veterinary practice to keep them in quarantine or apart, for about three weeks in case they might develop some disease they were incubating before you got them. Any coughing or running noses or uncontrolled shaking should immediately be reported to your vet for swift treatment. Do also report the symptoms to the breeder you bought them from. Call in the vet if one carries its head on one side. In most cases prompt action saves the day.

Moving and handling

"Pigs might fly if they had wings"
(English Proverb)

By handling pigs I do not mean stroking them, although both you and they will benefit from rubbing their backs and behind their ears, and stroking their tummies. This last action invariably results in the pig rolling over on its side. It is also a very good way to get a nervous, first time gilt to lie down on her side so that you can put her piglets to her. Pigs are gregarious and like company. They are intelligent and like to be spoken to and given a scratch.

By handling, I mean catching, moving and loading them. Tiny piglets can be carried by one back leg, two in each hand. It is the correct way to carry little piglets that oddly do not squeal once held so. They always squeal their heads off if grasped around the body.

Catching pigs requires being taught the correct way. Then the skill comes with practice. Tiny piglets are easily caught but bigger ones, up to eight weeks old or so, require a definite method. Drive the piglets into a corner. They will turn and look at you. Select the one you wish to catch and move in. Bend down and place your hands on either side of the piglet as it moves towards you, through your legs. Allow your hands to slip along its flanks to the point in front of its hocks, bring your thumbs round above the hocks and grasp tightly, lifting the piglet off the ground. This movement happens in just one second. It does, as I say, require practice.

Bigger pigs have to be moved on foot. Remember all pigs are greedy and are best 'towed' rather than driven, so move them when they are hungry with the feed bucket in front of them. Remember too, that if a pig sees a gap big enough to push its snout into, it will push to get through; so thorough preparation is a sound plan. Blank off all gaps, if possible, and use a pig board up behind them. A pig board is easily made from a two foot square of plywood with a hand hold cut in the top edge.

When moving store pigs along in the open you will require two people at least; one in front with the bucket of food and the pig board, and one behind, also with his pig board. The person behind should allow the pigs to go along at a steady pace and should not, on any occasion, get too close behind as this will make them apprehensive and they will turn and break back past him. One must be ready for this at all times for there is always one pig that will be dead set on doing just that. Concentrate on the job in hand and don't be distracted.

When it comes to loading adult pigs, bribing with food is always best tried first. An adult sow is heavy and a lot stronger than you are, so brute force should only be used as the very last resort. Some pigs are undoubtedly more stubborn than others. To get a stubborn or blinkered pig (Lop-eared pig) through a gateway or to go through a gap where you have removed a strand of electric fence, straw down a pathway through, and give her time.

To load a sow into a lorry or trailer, one must wait until the sow has her front feet well up the ramp before getting behind her to hold her there. Do not push her from behind at this stage. Bribe her with apples or food put down in a trail in front of her. Sometimes, with some encouragement, she will go in just like that, but not always. Sometimes a sharp slap on her rump helps. You'll find too, that this is the time she pees down your boot and steps on your toe!

If bribery fails, you can get a plastic bucket over her head and back her up into the trailer. This sounds alright but you need two strong men on the bucket, who know what they are doing.

Sometimes you need to pack up and move to a different location. Infuriatingly she may then trot in with no trouble at all, just when you'd decided to leave it, starve her for a day and then try again with half a bucket of apples.

Training

The best method is to train your pigs. Park the trailer in their paddock and feed them in it. There are a few more tricks of the trade you can try. To get a pig to go forward up the ramp, insert your fingers into the sow's hair, palm upwards, just above the tail. Now by squeezing your hand together, you pull the hair towards her head, in other words in the direction you want her to go. Do not pull too hard or she might sit down, but a quick tweak gets her going. It saves you standing there endlessly slapping her rump with the result that your hand is bright red and hurting, and she is more determined than ever not to budge.

Here's another trick to use on a stubborn sow that will not load but stands like a rock at the bottom of the ramp. You need three of you for this method. Take a piece of thin rope and slip a noose around one of her back legs, just below her hock. Now one or two of you pull with all your strength backwards, that is to say away from the trailer. The old sow immediately pits her strength against yours. When she's pulling with all her might, suddenly let go of the rope, so that she stumbles forward off balance, and the three of you quickly push her up the ramp into the trailer.

Here, a stick and a board are being used to control the movement of this Saddleback pig.
Photo: Katie Thear

Lastly here's a method to load store pigs or porkers that refuse to budge into a lorry for market. There is no need to beat them or shout. Make sure they can't get by or under the loading ramp with the lorry backed up to the race. Drive all the pigs away from the lorry to a shut gate or sheet of galvanised metal.When they arrive at the dead end they will turn. Now let one or two past you and when the others follow drive all of them at speed up and into the lorry. It does work.

Fighting

Sows will tell you by their squealings when one is in the wrong paddock, knocking eight bells out of its neighbour. If this happens don't panic! With a length of 2cm (0.75in) polythene water pipe take careful aim and strike the aggressor on the snout. One sharp blow, not too hard, will part them without damaging them and stop the fight.

A boar fight is altogether a different matter and all precautions must be taken to keep boars well apart in the first place. It would take two experienced people with pig boards and sticks, and great care or bravado to part them.

Clean, fresh water is essential and an automatically refilling system is ideal.
Photo: The Cotswold Pig Company.

A Range of Feeders and Drinkers

Feeders such as this galvanised iron one need to be shallow for ease of access and to avoid tipping.

A feeding trough can be made from strong timber but is not as easy to clean and the pigs may gnaw it.

This container can be used as a pellet feeder or as a drinker for a small number of pigs

Galvanised metal, rubber and plastic are suitable for feeders and drinkers, but round ones may be tipped over unless placed in a supporting base such as a rubber tyre.

Feeding

"A starved pig is a great deal worse than none at all."
(William Cobbett - *Cottage Economy* - 1821)

Pigs in their natural state eat a little and often. This is not practical, so unless you use ad-lib feeders, feed them twice a day, morning and afternoon. You'll find that you will have to stick to the same time every day. If you are late the pigs will certainly let you know by squealing until you show up with their food. They have an innate and accurate sense of time. The amounts required at different stages are indicated below:

Rations per day

Normal sized breeds		*Miniature breeds*
Adult boars and sows	1.5kg-2kg (3-4lb)	1-1.5kg (2-3lb)
Lactating sows	2kg (4 lb) & 0.25 kg (0.5lb) per piglet	1-1.5kg (2-3lb) & 0.1kg (0.25lb) per piglet
Piglets 2-4 months	Up to 1.25kg (2.5lb)	0.5kg (1lb)
Piglets 4-5 months	1.75 kg (3.5-4lb)	0.75kg (1.5lb)

For little breeds such as the New Zealand Kunekune, one needs to reduce the amounts, as shown. The little Kunekune pigs can cope with a very high fibre diet and are true grazing pigs, being able to maintain condition on good quality pasture alone. Of course pregnant and lactating sows should always be fed at a higher rate as their nutritional needs are greater. It is not generally realised that all pigs will graze, given the opportunity, some more than others.

Compound feeds

Compound feeds are the most convenient to use. They are formulated as complete rations and contain all the necessary nutrients, including minerals and vitamins. They are available in different formulations according to the age and type of pig, and can be bought as pellets, cubes, nuts, rolls, or as meal. Examples are shown on page 42. Note that the 'ash' referred to is the mineral element of the feed.

Caution must be practised when feeding milk. It can cause diarrhoea because of the fat content. If it is kept in drums and allowed to go sour this overcomes this problem in pigs. The *lactobacilli* souring organisms act as natural probiotics, as is the case with yoghurt.

Examples of compound feeds

Grower pellets
Oil- 5.8%
Protein - 20%
Fibre - 3.9%
Ash - 5.6%

Grower/Fattening cubes
Oil - 4%
Protein - 17.4%
Fibre - 6%
Ash - 5.5%

Sow cubes (dry and lactating)
Oil - 4%
Protein - 16%
Fibre - 5.5%
Ash - 5.9%

Organic grower's cubes
(pork and bacon)
Oil - 3.2%
Protein - 16%
Fibre - 4.3%
Ash - 6%

Organic sow cubes
(dry and lactating)
Oil - 3%
Protein - 15%
Fibre - 5.6%
Ash - 5.6%

Pot-bellied pig cubes
(pet pigs)
Oil - 3.2%
Protein - 15%
Fibre - 5.6%
Ash - 5.4%

(Information by courtesy of W. & H. Marriage & Sons Ltd).

If you don't want to use commercial feed rations, be careful that the supplementary feeds are of sufficient quality for the pigs. The protein percentages of a variety of foodstuffs are shown below:

Percentage of protein in common foodstuffs

Sow Nuts	16% - 17%
Pasture	about 3%
Bread	up to 8% and high in carbohydrates
Potatoes	0.5% and high in starch.
Milk	3.5%

The feeding of fodder beet in the winter period is very beneficial. It is high in energy, but with little or no protein, must only be fed as a supplement. It replaces the energy that pigs would have got from the grass and is therefore particularly useful when you have had to take the sows off the paddocks, when it is too wet.

Feed one good sized beet per sow, fed whole scattered in the straw.

Weaning

I am a great believer in keeping this simple. I wean all my pigs including the little Kunekunes at eight weeks of age, which is the traditional weaning time. Some pigkeepers wean earlier for economic reasons. By the age of three weeks the piglets will be picking up their mother's supplementary food off the ground. There is no need for expensive creep pellets if weaning is at this late age.

If feeding proprietary nuts scatter them on the dry ground, not in the mud. Do not feed 'cobs' to the sow for they will be too big for the piglets to eat.

By eight weeks of age they will be competing with their mother at feeding time, eating a considerable amount as their mother's milk dries up. A sound reason for weaning at eight weeks is the fact that a pig's fertility cycle is three weeks, and she comes into season (hogging) from 4-7 days afterwards. She is following her natural rhythm of the three week cycle. If a gilt with her first litter has great demands on her, as with a large litter of say 11 or 12 piglets, she may 'milk off her back' and become excessively thin. It is important in this instance, to wean the piglets from her at five weeks. She will then come hogging on the sixth week. You may want to miss the first hogging and feed her up for three weeks to improve her condition and then put her to the boar. It may be prudent to feed the piglets, because of their young age with a higher protein feed such as one of the grower feeds shown on page 42.

To wean piglets it is best to put the sow and piglets where you intend to keep the weaners, to begin with. The next step, after one or two days, is to take the sow far enough away and secure, so that she cannot hear the piglets and they cannot hear her. Do not wean gradually, do it directly and abruptly. Believe me the sow and the piglets will not suffer.

Feeding the weaners

Feed them twice a day and not too much, too soon, or they will scour. Dietary scours can quickly turn to bacterial scours and must be treated before you start losing your weanlings. It is important to give the weaners access to clean water at all times. Remember that water pots or drinking bowls should be low enough for the piglets to reach.

Adult sows can drink well over 9 litres (2 gallons) a day and more in very hot weather and when in milk. A very good tub is made from an empty 205 litre (45 gallon) plastic drum. Cut off the end at 23cm (9in) deep with a hand saw. Now go around to your local tyre merchant and beg a couple of old lorry tyres. Simply drop the tubs into these. The internal diameter of the tyre needs to be about 60cm (24in).The tyre prevents the pigs from tipping the water tub over, something all pigs delight in doing.

You will need to place a concrete block into the tub of a sow with a litter to prevent the piglets from drowning. Little pigs like the Kunekune need shallower water tubs, bowls or troughs.

I cannot stress too much how important it is to check that your pigs have water at all times, especially during hot weather. Pigs can die or become very sick of salt poisoning if deprived of water.

Feeding the finishing pig

It is important, first of all, to know the amount of space required to accommodate various sizes of pigs. Too few pigs in a pen is a waste of space and in winter, when it is cold, they will lie very close together for warmth. A full pen will be warmer for them as they generate a considerable amount of heat. They will also make better use of their food and come to pork weight quicker than they would using half their food just to keep warm. This last point is one very good reason for finishing pigs indoors, but with access to an outside exercise run, rather than running them outside, where they may make very inefficient use of their feed.

An 18 kg (401b) weaner needs approximately 1858 sq.cm (2sq.ft) of lying space and a 75kg (1651b) baconer needs 5574 sq.cm (6 sq.ft). It is important that at all stages, pigs have enough trough space so that they can all feed at the same time. 36cm (14in) is the magical figure, enough for pigs right up to bacon weight.

I am not a believer in ad-lib feeding at any stage of the growing period. I much prefer to feed my pigs twice a day. Nor am I enthused by the feeding of wet meal. Feeding the meal dry gives the smaller pigs in the group a greater chance to eat their share, because the big ones cannot gulp it down, as they can with wet feeding. The old adage of giving them as much feed as they can clear up in twenty minutes is still, in my opinion, a very good one. If they scrabble and fall on their food and it's all gone in a few minutes, give them more. If they still have meal left after half an hour, clear up the uneaten food and cut back on the ration.

Start the weaners off on half sow nuts and half meal. I use one mix right through to finishing, and gradually phase out the nuts until the weaners are having 1-2 kgs (2.2-4.4 lb) of meal a day. Do not feed too much, too soon, for this may very well make them scour. If this happens cut the food back quickly and subsequently increase it more slowly.

Home mixed feed

The cheapest food is one that you mix yourself. You do not require expensive machinery to manufacture your own feed. All it requires is a clean shovel and a flat, clean area of floor. Buy the 'straights' from your merchant. Straights include barley meal, coarse ground wheat, fish meal and soya meal. A good standard mix would be:

10 parts coarse wheatmeal — 1 part soya meal
8 parts barley meal — 1 part fish meal
plus minerals

As the pigs get older, three quarters to pork weight, one should drop the fish meal and double the soya meal. Towards the end of the finishing, drop the soya meal back to one part and increase the barley meal by one part. Learn to feed by eye, that is one of the skills of a good stockman.

Household scraps should be avoided on health grounds. Since the major outbreak of Foot and Mouth disease in Britain in 2001, the feeding of swill is banned.

Grazing

All pigs graze, some more than others, and some breeds eat grass a lot more readily than others. Grazers tend to be the short-nosed, 'improved' breeds. Old breeds such as the Tamworth have long noses adapted for rooting and foraging under trees, rather than for grazing. All wild or feral pigs: the wild boar, the Bulgarian native, the Tibetan, the Captain Cooker from New Zealand, and several African breeds, have long noses and are similarly adapted as woodland pigs.

All pigs will root to some extent, especially when grass has disappeared, and this is damaging to pasture. One can stop them digging to a degree by putting a ring in the nose. I dislike the practice and never ring my pigs. I feel that one is denying them of one of their basic senses. It is as bad as keeping breeding pigs all their lives on concrete so that they never get the chance to root in God's good earth.

When the grazing is plentiful dry sows can certainly exist on grass alone, but with unborn piglets developing inside them, they need a certain amount of extra protein in order to produce viable piglets. The growing store pig also needs more protein than the grass can provide.

The protein content of pasture rapidly drops off as summer approaches and the grass runs to seed. I tend to disregard the value of the grass and treat it as a beneficial 'filler' that counteracts constipation.

There is no doubt that if you farrow your sows indoors both the sow and her litter do so much better outside in the sunshine, so put them out as soon as they are strong enough, or the weather allows. This is usually from about a week old, but they must have good shelter and warm straw.

In spring, when the weather can be so fickle, it is a good plan to divide up the field or grass area so that the pigs' grazing is rotated. (See page 23).This rests the paddocks, allowing them to recover and not become 'pig sick'. In the winter it is wise to have an area as a 'sacrifice' paddock. That is to say, you keep the pigs here all winter and if they plough it up and tread or poach it badly you don't worry but reseed it in the spring. The only cultivation required would be to drag a harrow over it, broadcast some grass seed with a liberal amount of clover in the mixture, and roll it in with the heavy roller. Do have clover in the mixture; it takes an awful lot to kill it and its persistence means the paddocks will always be green. The pigs' manure certainly puts a lot back into the soil. You will find that there is no need to put any artificial fertilizer on the paddocks.

Pigs and docks unfortunately go together. This pernicious weed grows only in good soil of high fertility. The only sure way to kill off the docks is to spray them with a specific weedkiller, but many people are unhappy at doing this from the environmental point of view. Continual mowing weakens docks but won't eradicate them. You can of course pen your pigs very tightly on the offending areas that are festooned with docks, making the pigs root them out. You will though, still have to pick up all the unearthed roots and burn them.

The folding of pigs on arable crops

The folding of pigs on arable crops is really only suitable for adult pigs and then only for dry sows. Sows with litters are not suited because of the difficulty of confining the piglets and also the mud in wet weather. Only well drained, light land is suitable. Pasture or grassland is, of course, suitable for farrowing sows and their piglets, in suitable arks.

Arks or huts on skids, which are easily moved and with the aid of electric fences make the folding of dry sows on arable crops comparatively easy, provided one cuts or mows down the standing crop where the electric fence is to be placed. Long narrow paddocks for the sows are best, so that they have a narrow feeding face for complete grazing and with the hut at the other end the sows get plenty of exercise. Concentrates in the form of cobs are fed in a long line, ensuring all the sows get their share. If sowing a grass ley for them, as referred to earlier, include clover in the seed mixture. It is so beneficial in the pig paddock, because in my experience, pigs cannot kill it and even if they turn the turf over, it will grow again.

Fodder crops

Historically, pigs were killed for meat in the autumn because there was not enough food to keep them through the winter. Then came the turnip which was sown in July - August and harvested in the autumn, enabling farmers to keep their animals through the winter.

There are a range of crops that can be grown for autumn and winter feeding, including potatoes (feed cooked), Jerusalem artichokes, carrots, fodder beet, parsnips, turnips, swedes and mangolds. As a rule, the bulky items include a high proportion of water and so have less feeding value by unit weight. Those with a higher dry matter content are most beneficial.

The green foods like cabbage, kale, legumes, maize and grasses can also be useful as forage crops. Roots fed with skim milk (if available) can be a useful addition to the diet.

Outdoor piglets do not need iron injections because they get minerals from the ground.
Photo: Katie Thear

Breeding

"The boar is half your herd"
(Traditional saying)

The boar

The young boar should be kept in a small paddock so that he may take exercise. If he is kept in a pig sty let him out for a time every day, to build up muscle in his legs. The space for the boar must be big enough to give him room to exercise and to serve his gilts and sows. It is recommended to give him a total area, including lying area of 10 sq.m (108 sq.ft). One should site the boar pen so that he can see and hear other pigs. It is a great advantage to have a sow pen next door to him. The walls of his pen should be 1.5m (approximately 5ft) high for safety.

Young boar

The boar's pen and the sow's next door must be insulated and/or shaded in such a way that they remain cool during the heat of the day in summer. He may also be allowed to run with his sows all the time, but one needs to be observant and make careful note of the dates of service and whether the sow returns to service after three weeks. Good records need to be maintained accurately and diligently, detailing dates of vaccinations, service by the boar, farrowing and weaning dates.

The protein in the diet of the boar should be increased prior to using him at the earliest age of eight months but preferably at ten. It is essential to ensure boars are not required to work before they attain sufficient size, weight and are old enough. The boar should not be over-fat to work but be big enough so as not to be discouraged from mounting his gilts. It also helps to ensure that, once mounted, a successful entry (lined up with your knee and the sow's tail) and service are achieved.

To have a young boar looking up at a hopeful gilt or a belligerent old sow, won't do a great deal for his libido or ego. I emphasise that you must make sure your boar has been fed on a high enough plane of nutrition and has an adequate chance of a good first service, for the preparation of a long and productive life in the herd.

Sow

Remember always take the sow or gilt to the boar, never the other way round. The boar will be too distracted by a new sty or paddock and will have to 'mark' it up as his own, before he can concentrate on the gilt.

Do not let gilts and especially, sows get over-fat. Start serving gilts when they are ten to twelve months old; if left to get too old and fat there may be difficulties in getting them in-pig. Fat sows lay down a lot of internal fat around the ovaries which inhibits ovulation. Miniature pigs may be ready for service at six months old. Do not let the boar get overweight either, as it puts a great strain on his feet and pasterns. Never buy a boar that is down on his heels; he should stand up on his toes. Buy the best boar you can afford, bearing in mind that 'the boar is half your herd', meaning he will have the greatest influence on the sows' offspring and production. Ask to see his father and make sure he was one of a large litter, and that this was also true of his father before him.

Never overwork a young boar. One or two gilts a week is enough. If you have only a couple of sows it is better to use someone else's boar and take your sow to him. Do not expect him to hire you the boar and be suspicious if he says you can borrow one. He's probably been all over the country and heaven only knows his health status. I once had a pig breeder ring me up here in Dorset from Surrey. She wanted to know if I had her boar. *"No"*, I said, *"Have you lost one?" "Yes he's somewhere on the South Coast"*, she replied.

Artificial Insemination (AI)

AI is increasingly being used by small pigkeepers who do not necessarily want the responsibility and expense of keeping a boar. A semen delivery service will deliver semen by post the day after a telephone order is received. All that is necessary after opening the insulated delivery container, is to have ready a disposable catheter, some non-spermicidal jelly and some paper towels. As long as the female is still on heat, it is easy to insert the catheter into the vulva and inject the semen. She will stand rock-still with an arched back while this is going on. The only problem likely to arise is if she has gone off heat in the intervening period before requesting and receiving the semen. One can buy aerosol cans of 'boar odour' to trigger off heat.

AI offers semen from some of the best breeding stock, and its use does mean that movements of breeding stock on and off the farm are not required. This reduces the risk of introducing disease.

Hogging

Sometimes it is difficult to tell when an old sow is hogging (on heat), although when the piglets are weaned off her, she will come in from four to eight days after. One service, when the sow is hogging well, and one more 18-24 hours later is all that is necessary. The later the service heat period, the more likely a big litter.

Gilts are relatively easy to tell when hogging. They usually have a reddened and swollen vulva, and perhaps a little mucous. Look for a change in behaviour. She may shun her food or become pushy or aggressive, and be more vocal. If the boar is near by he might well 'bark'. It is always a good idea, if it can be arranged, to pen the sow near the boar. His pheromones will stimulate her. If you dig your knee into the sow's side and then sit or press down on her back she will stand stock still, and if you are very observant you will notice that she twitches her ears.

A sow on heat will stand like a rock, refusing to move.
Photo: Katie Thear

Be on hand to make sure that the boar is gentle and correctly lined up with the first time gilt and conversely that the old sow is gentle with a virgin boar. You would think that it would all be so easy to put a sow to a boar. You've weaned the piglets, waited four or five days, rung up the boar owner that you are coming with your sow. All you have to do now is load up the sow and be on your way.

In the end you ring the boar owner to tell him that you have been trying all day to get Rosie in the old horse box, but to no avail. It happens one way or another to us all. A few years' ago I rang a British Lop breeder I know, to ask if I might bring my sow down to Honiton to one of his boars. I made sure she was hogging when we left home to travel the fifty odd miles to the farm. When we got there she had worked herself up into quite a stew. The owner of the boar explained he didn't have anywhere to put her. So he let his young boar out to serve her in the yard. Well, our crabby old sow flew at him drawing blood and beating him up. With pig boards we parted them and decided on our next move. In the end the owner said, *"Take the boar home"*.

We got the sow back in the trailer, put the partition up behind her and walked the boar in. We set off home but half a mile from the farm, going back up the steep hill, there was a tremendous crash. I stopped immediately to find the boar upside down on his back with the broken partition and an irate sow on top. We got the sow out with difficulty where she stood immobile in the road utterly fuming with rage. While my wife stayed with her I took the boar back. *"Don't worry,"* said the breeder,*"We could put her in the paddock with the old dry sows and the senior boar."* The young boar was put back in his pen and I returned up the hill to find our bad tempered sow, still foaming at the mouth and rooted to the spot. Once she realised that he was not in the trailer she went back in. I had the trailer facing home. By driving round in a loop around the lanes we got back to the paddock where the senior boar was. He was asleep under a hedge and absolutely enormous. We tipped her out and returned home. The boar's owner rang exactly three weeks to the day to say she had been served and had been as good as gold!

I relate this story to illustrate how a longish journey in a trailer will either bring a sow to hogging, or if she was hogging like mad when you left, may have completely gone off the idea when you arrive at the boar owner's farm.

Caring for the in-pig gilt

Now the gilt is in pig. She has gone the three weeks and not returned to hogging. Do not overfeed her and get her too fat. She should not be so thin that she slopes away either side of her backbone, but needs a level back, 'fit not fat'. Put her food out in a line of dollops so that she has to walk for it. Increase her ration only in the last three weeks to double the amount. Put her on her own in a small paddock, preferably where you can keep an eye on her. If farrowing outside, or in a farrowing pen, put her in there a few days before she is due, to give her time to settle in.

Farrowing

She can farrow outside in an ark, provided it is not baking hot. In winter the piglets need a heat lamp suspended above the creep area. If it is exceptionally cold and the farrowing pen is a converted loose box, suspend a sheet of plywood, three feet off the ground above the farrowing area. This raises the temperature a little more as it tends to keep the lamp's heat in.

When you separate her for farrowing, treat her against worms and external parasites. *Ivomec* or the equivalent *Dectomax*, are very good as they do both, and also give the piglets immunity for three weeks or so, through the placenta.

If farrowing outside in an ark put no more than half a small bale of straw in it. Too much and the newly born little piglets might get trapped in the straw and be laid upon. If your farrowing ark has a door, shut it just before and during farrowing to keep the piglets warm and to give the sow some privacy.

The udder will develop in the gilt up to a month before she is due, and will harden twenty four hours before with milk before she actually gives birth. Test for milk by squeezing a teat at the base and drawing down with the thumb and forefinger to eject proper milk. Sometimes one can squeeze out a watery substance, perhaps two days before.

She will start 'nesting' two or three days before by collecting litter, grass, dock stems and tearing up anything she can find. I had an old sow who chewed up a hundred metres of polythene water pipe to make her nest!

General signs of approaching birth are restlessness with a change of texture of the udder from being loose and soft to tight and hard. The vulva will get pink and swollen. If the udder is very hot and red and obviously painful to her, suspect mastitis.There will be a slight ridge along the length of her udder, between it and her abdomen. Mastitis occurs more in the summer with sows fed too well, ie, rich grass and too much concentrates. *Quarterzone* injection will reduce the swelling. If she does succumb to it, do not take the piglets away from her, for they stimulate her to produce milk, although you might have to resort to treating her three or four times a day with *Oxytocin* if her milk dries up completely. If this happens top up the piglets with a bottle. Human baby milk powder is an excellent substitute but it

A heat lamp set up on the other side of the rail where the sows cannot gain access.
Photo: Katie Thear

must be diluted. Tinned *Carnation* condensed milk will do as a stop-gap. You can put the milk into a shallow dish and within 24 hours the piglets will lap it up; they're not stupid.

To test for farrowing time give her some straw; if she rakes it back with her fore feet as opposed to taking it in her mouth, she'll soon be on her way. Pigs do not produce a water-bag like cows or sheep. Immediately before a piglet is born the sow will thrash her tail. Do not interfere, stay calm and quiet. As the piglets appear wipe them off with a towel to dry them as this stops them getting cold, and put them to a teat or under the heat lamp if it is very cold weather and if the piglets are not strong enough to suckle. The sow may well get up and down between piglets being born; this is quite normal.

The womb is in two parts, rather like thumb and finger of a glove and often she needs to get up and turn over to give birth to those on the other side. Sows take from 1½ to 8 hours to farrow and it is quite common for piglets to be born facing backwards. These will have a great deal of mucous in their mouths. Fish it out with a finger and make sure breathing is not impaired.

If the piglet is not breathing check to see if there is a pulse at the base of the umbilical cord. If there is, do not give mouth to mouth resuscition as this will blow the mucous straight into the lungs and the piglet might die. The trick is to hold the piglet upside down for five minutes or until it starts breathing. Use mouth to mouth resuscitation only as a last resort. Do not swing the piglet round and round, as one does with a lamb. New born piglets are incredibly slippery, you might have the piglet slip out of your grasp and land, hung up in the rafters of the farrowing pen.

Towards the end of farrowing, piglets are often born amongst the afterbirth and sometimes a piglet may get trapped. If left it may die, particularly if it is a weak one. Stillborn piglets are not unusual. The sow may well start to shake and shiver during farrowing. This is quite normal; she is not cold.

If the sow is very restless during farrowing it may be necessary to take the piglets away as they are born, and put them in a cardboard box with straw. This is a temporary measure for their safety and until she settles down. Alternatively you can put them under the heat lamp in the creep and keep them there.

Some people will tell you to give the sow beer to calm her down. It doesn't work. If the sow is that restless you certainly won't get her to drink anything, let alone a pint of beer. If she really is nasty and starts going for her piglets and you as well, give her an injection of *Stresnil*; 5ml. no more, and less for a small breed.

You will know when a sow is near to finishing as she will grunt happily and rhythmically as her litter suckles. Sometimes you get a 'squeaker', one that will not feed and its squeals upset the mother. The thing to do is immerse the piglet up to its ears in a bucket of warm water for a while, then towel it off. This normally does the trick. This is also the best method of very quickly warming up a piglet that has crept away into the cold and wet on a frosty night, when its mother is busy farrowing and suckling its brothers and sisters. Again towel it off and put under the lamp until its strength returns and you can put it to a teat.

Almost immediately the last piglet is born the sow should expel the afterbirth. If after a few hours nothing has happened, get a vet or if you are proficient yourself, give her some long acting *Penicillin* and about 4ml of *Oxytocin* without delay, before she goes off her food. On rare occasions, and more especially with the older sow, she will stop and give up straining altogether, particularly in very hot weather, after just two or three piglets.

After scrubbing your hand and soaping your arm well, it will be found on internal inspection that there is a dead piglet blocking the way. The womenfolk are better at this, as they have smaller hands and arms. If you have put your arm in up to your armpit and cannot feel anything, give a 1ml injection of *Oxytocin.* This will start things off again. You might have to give 1ml every half an hour to keep things going with old sows. They lose the muscle power to farrow a large litter and if left, eventually have a large number of dead piglets.

If after an internal examination the vulva is bruised and swollen, don't worry, it will soon subside in a few days but it is essential to give the sow a course of *Penicillin*, or at least one injection of long acting *Penicillin*.

Remove the afterbirth and spread a little clean straw. Allow the sow to suckle her piglets undisturbed. They will drink every 20 minutes for the first 48 hours. Leave her alone to rest. Give her some water to drink, preferably warm in very cold weather, and some food six to eight hours later with fodder beet or bran if she has

The sooner they start suckling, the better. *Photo: Graphics Collection.*

a tendency to constipation. Do not give her the same amount of food as she was having before she farrowed. Feed her half the quantity for the first two days, otherwise you will overface her and put her off her food with the subsequent result of lowering her milk yield.

If the piglets' cords are very long, over 30cm (1 ft), one can simply break them off at six inches, but not too short so that infection can enter. There is usually no need for iodine; piglets' cords dry up in a few hours and always drop off in three days.

Little pigs can be very adventurous and care must be taken to ensure that they can get back into the ark again (if farrowing outside) by providing a suitable ramp of earth. Fit a fender to the opening of the farrowing ark if you have one; it stops the piglets straying and cuts down the draughts at low level in the ark.

The sow's food should be increased gradually over the next few days, not all of a sudden which might put her off her feed. The old formula was three to four pounds for the sow plus half a pound, increasing to one pound for every piglet. As referred to earlier, small breeds like the New Zealand Kunekune will need half this amount.

It will be when the piglets are two to three weeks old that the sow should be having up to the maximum ration. At this time the piglets are making big demands on her milk supply. By the time the piglets are six weeks old they will be eating a considerable amount of their mothers' feed, so make sure you are feeding enough nuts for them all.

You will be told lots of stories, including those about clipping the piglets' teeth, something we don't practise nowadays. Iron injections for the piglets are unnecessary if pigs run out of doors, for the sows will pick up ample iron from the soil. Pigs housed indoors all the time are a different matter and will benefit from iron injections, although the old men used to pull up a sod of earth and throw that in the corner of the creep for the piglets.

The farrowing of the pigman's gilts and sows is always a thrill to him, even the real old hands. Do not be put off by the disaster stories.

The new-born piglet and its care

Much of the mortality in newborn piglets is due to lack of nutrient liquids and hypothermia. Weakness through dehydration and chilling make them susceptible to coughing and starvation. The new-born piglet has no fat reserves and its digestive system is under-developed. There may be up to 7 or 8 hours between the first and last piglet being born in a litter, and time has no bearing on weight, meaning the last piglet may not be the smallest.

The colostrum quality may also differ between the teats of the sow and certainly the immunoglobulin levels fall during the farrowing period. If the small piglets in a large litter are amongst the last to be born,they will need assistance to survive, and enough energy to find a teat. A high protein and energy pig booster with iron and vitamins in a dispenser is what one needs. You pump one or two squirts directly down the piglet's throat. There are several piglet boosters on the market; they should have high protein and energy content of *Triglyceride* based on colostrum substitutes. One of these provides nutrients for the maintenance of body tissues during the first hours of life.The high concentration of antibodies and naturally occuring beneficial bacteria restrict the development of the pathogenic bacteria in the gut. The energy will help to avoid hypothermia and also help the piglet to optimise the value of the sows' colostrum.

Pedigree breeding using line breeding

One of the most satisfying and fascinating sides of pig farming is the breeding of pedigree stock. If done diligently it can be most profitable as your aim will be to improve your pigs within a chosen breed to perfection of that breed's standard. Remember that good stock always sells even in hard times.

If two pigs have a common ancestor it will be found that the chances will be increased that their genes will be the same. The closer the relationship, the smaller the fraction of their genes that are probably dissimilar and vice-versa.

Outdoor piglets are less likely to suffer from infection than those raised inside.
Photo: David Thear

The more distant the ancestors, the less chance they have of contributing any particular genes to their descendants. A parent always contributes half of its gene complex to its offspring. A grandparent contributes about one quarter, on average, and so on. Every generation, which comes between any one ancestor and its descendants, reduces by half the proportion of genes, which are likely to be exactly duplicated in the descendants.

The method of measuring relationships can be most simply illustrated in the following diagram:

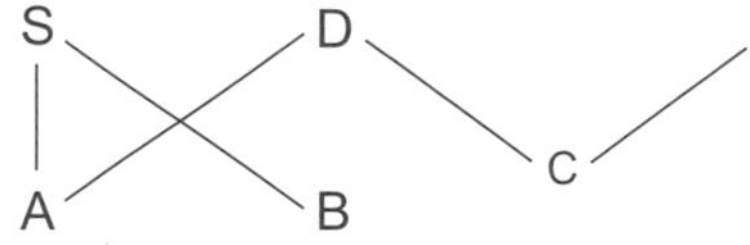

A and B are brothers by sire S out of dam D, and C is a half brother to A and B, being out of the dam D by another sire. There is a direct relationship, parents and offspring (S to A and B; D to A, B and C), the latter being descendants of the former. There is also a collateral relationship between A, B and C, being 50% and 100% between A and B of 'common blood'.

Line breeding tends to set up separate 'families' and to isolate them from the rest of the breed, and as it gives scope for selection for production characters, it is particularly suited to those situations where the more complicated gene interactions are involved. In line breeding (as opposed to in-breeding) it gives greater freedom of choice among ancestors and collateral relatives and allows also a greater opportunity of exerting selection for other characters than pedigree alone. The differences in performance between distinct 'families' resulting from line breeding, therefore gives better opportunities for picking out the different genotypes successfully.

The quickest way to increase the frequency of desirable features or traits, with culling of the undesirable, is to line breed. The object of line breeding is to keep the offspring more or less closely related to a particular ancestor, which has the desired characteristics. Where such desirable combinations do occur the chances of retaining them in your pigs are increased by line breeding to them, and only by line breeding. Once you start out-breeding you will immediately lose all that you have strived for to attain your aim. The matings will produce pigs you won't want and certainly disrupt your breeding programme.

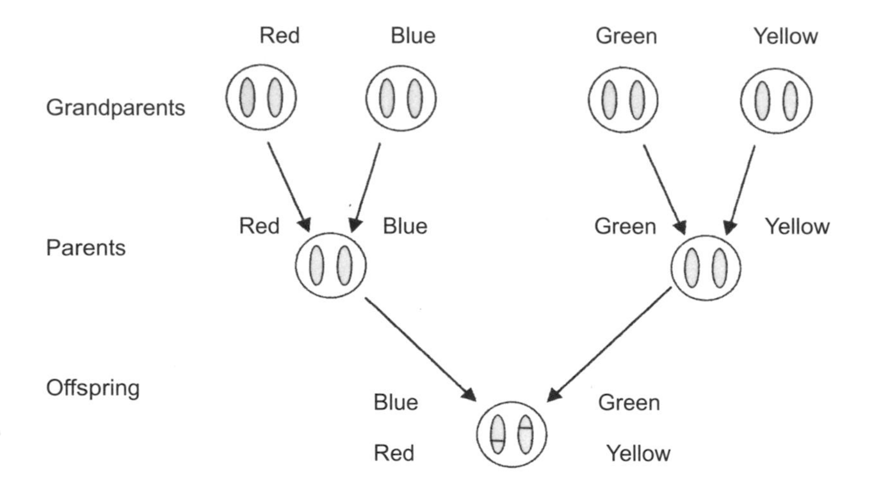

This illustrates that, although offspring get exactly half their genes from each parent, the contributions from grandparents can vary. On average a quarter of an animal's genes will come from each grandparent, but by chance, individual animals may receive greater or smaller contributions from particular grandparents. To sum up, I quote a summary paragraph on line breeding used for some *Rare Breed Survival Trust* notes, sometime ago:

'*Line breeding combines mild in-breeding with selection; its purpose is to avoid diluting genes from valuable ancestors. Because the genotypes are relatively well known and only related animals are used in the mating programme, in-breeding depression can be mimised. The technique avoids losing desirable characteristics, which might occur if unknown breeding stock were introduced to the population. A particular advantage of the technique is that favourable gene interactions, if present, tend to be preserved".* Hence the old maxim: *"Breed tight and cull hard".*

Health

"Experience keeps a hard school,
but fools refuse to learn in any other way".
(Traditional saying)

One should always have in mind a picture of what a healthy pig should look like, so you can judge it against one that is ailing or merely off colour. A healthy pig always has a curly tail, although the New Zealand Kunekune does not curl its tail up tightly. It should always have a silky coat and clean skin, free from redness and roughness. Lastly it should have bright eyes and be alert and aware of you and its surroundings. It should always come to you at feeding time with a good appetite. It may come to its feed to start with, poke it about and then lie down again, so observe the pen of pigs for a moment or two. If your pig falls short of any of these pointers then take steps to find out what is wrong. It is true that often when one of your pigs is off colour it can be difficult to find out why.

A layer of fat makes it virtually impossible to hear the actions of the heart and lungs. It also often prevents one from feeling any internal organs to find the fault therein. These are things you can do before you call your vet:

• Take the pig's temperature with a lubricated thermometer that is inserted into its rectum. The normal temperature is 38.6^{0}C - 38.8^{0}C (101.5^{0}F - 102^{0}F). A rise in temperature denotes a fever. Antibiotics from the vet will usually lower it.

A sub-normal temperature is not good. Often it is already too late but do confine her to a small area, pile her with straw, hot water bottles and blankets or suspend a heat lamp over her.

• Examine her nose and vulva for discharge. Tell the vet if there is any.

• Count the breathing rate. It should be 20 to 30 per minute (Piglets 50 per minute).

• Put your ear to the pig's stomach and listen for any gurgles or lack of any movement in there. If you can feel a pulse on the inside of the front or back leg, count the beats. A pig's heart pulse is 70 to 80 per minute.

• Look at the mouth; are the gums pink or blue?

• Look at the whites of her eyes; are they yellow?

If the cause or illness is not obvious ring your vet immediately. Report all these points over the telephone, giving as much information as you can.

If two or three pigs are ill at the same time it is essential to call the vet. For instance, if piglets start snuffling or coughing it may mean viral pneumonia and if treated early is easily cured with penicillin. Little piglets have no resistance to disease and need treating very quickly. Old sows can be much more resiliant.

Summer infertility and heat stress

It's essential to maintain a wallow and shade for every paddock. Pigs like to be almost submerged in a wallow. Mud is better than clean water as it retains more moisture on the body. Trees are best for shade but lacking these, pig shades are easily put up. Old galvanised sheets rigged up on 4 posts 1.5m (5ft) off the ground, work well enough if funds are limited. They are marvellous things, old sheets of galvanised iron; a pig farmer should never be without them. They are useful for funnelling the pigs when loading and as makeshift housing or a crush when injecting weaners and, of course, for filling up gaps in the fencing to stop the little darlings getting out. The odd sheet in the hedge must surely characterise the pig farmer's property. These same corrugated sheets make good "troughs" for feeding sows outside in muddy paddocks during winter. Put the sheets down on the ground and throw the pig nuts or cobs onto them. If they get curled up by the weight of the pigs, just turn them over and stamp them flat again.

One very useful tip if using arks with wooden ends, is to cut a window 45 x 30cm (18in x 12in) in the back end with a chainsaw, in the summer so the pigs within have a through draught. You can easily cover the hole with a piece of plywood, screwed on, when the weather gets cold.

Boars present particular problems in the summer. Summer infertility is caused by heat stress, which doesn't show up until the end of autumn. Sperm production shuts down if the temperature of the testicles reaches 41^0C(106^0F). Boar semen takes 34 days to produce and ten days to mature before ejaculation. There is a 15 day tag between suffering heat stress and becoming infertile, so if your boar gets overheated it will be 60 days before semen production is back to normal. The situation is made worse by the fact that from mid-summer, hormone changes are induced by the shortening of the day's length. This reduces the boar's sperm count.

Sunburn or sunstroke should not be confused with heat stroke. Sunburn is more common in piglets under 16 weeks of age. The symptoms are characteristic. There is a redness of the skin, which becomes blistered. They move with a characteristic staggering action and suddenly hollow their backs as if stretching. They should be moved to shade immediately and kept cool. Apply calamine lotion, never pig oil, as this tends to fry them in the sun. Cloths dipped in cold water and held behind the ears will lower their temperature.

If farrowing outside in summer it is important to have a wallow and shade near the farrowing ark so the sow is close to her piglets. Put straw bales up and over the roof of the ark and tie them tightly together. This dramatically lowers the temperature in the tin-roofed ark.

You need to keep a close eye on the farrowing paddock when it's hot because the sow is more likely to farrow outside her ark. Hot weather makes supplying just the right amount of straw in the farrowing ark quite critical. Too much and she'll

The pig man's medicine cupboard

One needs to keep a few instruments, medicines and dressings, for emergencies, as follows:

- A thermometer
- A pair of round ended scissors
- 10ml *Arplex* plastic syringe (reusable)
- Needles - 17 gauge x ½" - Luer type to fit
- Also 17 gauge x ¾" or 1"
- Roll of cotton wool
- Calamine lotion
- Pig oil
- Liquid paraffin
- Surgical spirit
- Bottle of multivitamin injection
- Bottle of Vitamin B injection
- Bottle of lactulose
- Bar of *Ex-lax* chocolate
- Erysipelas vaccine
- *Ivomec* or *Dectomax* wormer injection
- Wound powder and antiseptic cream
- Crude disinfectant (*Jeyes Fluid*)

Also:

- Ear notching pliers
- Ear tattooing set
- Nose rings and nose ring pliers, if ringing is to take place

be too hot. Too little and she won't be able to nest, and will farrow outside, with the resulting loss of piglets.

Sows that are too fat, in fact all pigs that are carrying too much condition, cannot cope with heat as well as their leaner brothers and sisters; this is because their core temperature rises more relentlessly.

Signs of heat stress and how to cope with it

The pig will be lying on its side, panting. If not treated, the panting will increase in timing and she will have difficulty standing up and worse still, be unable to stand at all. Lastly she will become unconscious with resultant death. It is therefore vital to lower her temperature as quickly as possible with a hosepipe. If this is not practical, use wet hessian sacks, towels or anything that can be kept wet with buckets of cold water. Ice cubes pushed up her anus and held behind her ears work well, in addition to the cold water. The most likely candidate is the overweight sow in the throes of farrowing. Stick at the job, it is not a quick process; it can easily last several hours. Keep monitoring her temperature with your thermometer.

At pig shows the penning is often made of solid sheet metal gates. These can become quite hot very quickly rendering the pen where your prize pig dozes, not unlike an oven. We always take a bottle of vinegar with us to agricultural shows to apply just behind her ears. It helps to lower her temperature; as it evaporates it cools the same way as surgical spirit does. Do give your pigs plenty of water, enough to wallow in. Wet the bedding if the show pens are put up on concrete, and always take some shade netting to put up over the pig pens.

A tip worth remembering, if you find one of your pigs with heat stress, is to quickly dig a wallow beside the suffering pig, line the pit with a sheet of polythene (or a plastic sack opened out) and then fill with water. Now roll the pig over into the wallow; this will greatly help to reduce its temperature. This method will be no use at all to a sow that is farrowing on a boiling hot day, of course. If she is in an ark and you cannot get at her, lift the ark right off her. Rig up some shade and persist with the cold water and ice cubes. It is much better to avoid heat stress, by bringing her into a cool building *before* farrowing starts. Old farms are often blessed with old stables and barns that are always cool. Not everyone has such a building, so we must make do with what we have and insulate the roof.

Scouring

Scouring piglets of any age need to be treated quickly. There are proprietary preparations on the market; they come in containers that pump a squirt straight down the throat of the sick piglet. The best cure I have found is the use of *Tylan* power, diluted and then added to the drinking water. It acts like magic on scours in weaners and is also very good at nipping in the bud any respiratory problems.

Scouring in little pigs is invariably due to a rapid change of diet or over-feeding and also keeping the recently weaned pigs in cold, draughty conditions. Treat all scouring piglets quickly; dietary scours can easily lead to bacterial scours that kills piglets very swiftly, often whole litters. Sometimes you won't see very loose dung because the scour is just like water; so watch for wet tails, it'll be the only sign you'll get. Good hygiene is essential. Hose down and disinfect farrowing pens and clean out weaner pens every day.

If you find a little piglet dead suddenly, suspect E. coli immediately. It's a real killer. Get veterinary help straightaway. It's possible to vaccinate your sows two weeks before they farrow to prevent it but it's probably not applicable to the small pig keeper. The prevention of a build up of disease is one very good reason to farrow sows outside in purpose built farrowing arks.

Meningitis

Meningitis is the inflammation of the membranes that cover the brain. There are several forms of meningitis, one form starting with a middle ear infection. It may also be found as one the symptoms of erysipelas or tuberculosis and very often as a result of stress, often caused in my experience, by sunstroke, when up to a week later meningitis may appear.

The first symptom is restlessness and then often frenzied activity for a short time, followed by listlessness and then the pig will burrow in the straw. It will not eat, and will have difficulty in getting up and walking about. The animal may become blind, and there is usually constipation and retention of urine. The disease is usually fatal.

Treatment is not very satisfactory. Call your vet immediately and think hard about possible causes before he arrives. Was the pig stressed in any way before? Has it been overheated in a hot sty on a stuffy day or at an agricultural pig show? Has she been loaded or transported in such a way as to cause her stress? Sometimes though, you get no warning. On entering the finishing house you find a porker dead, or perhaps one thrashing about violently, kicking or paddling its legs as it lies on its side. Usually in a very short time it dies. If you can get some antibiotics into it very quickly, it may recover and when fit enough it is wise to send it off for slaughter as soon practical.

Erysipelas

The Erysipelas bacterium, *Erysipelothrix rhusiopathiae* can live outside the body of the pig, in the ground, for many months and resists drying out, which is why it was mistakenly thought of as a summer disease. It is for this reason alone that one should vaccinate against it.

Erysipelas used to be one of the principal diseases of the pig. However serious losses can be avoided by sound hygiene, vermin control and the correct use of vaccination. The disease manifests itself in the pig in a number of ways:

Hyperacute: Pigs are found dead, usually with purpling of the ears, belly and other extremities. It is most common in growing pigs over three months, although it is occasionally seen in adults. Hyperacute disease may appear in isolation or be a component of an outbreak.

Acute: This is the most usual manifestation, producing a lethargic pig with little appetite and a high temperature, 41-42^{0}C (106-108^{0}F). The pig will have an increased respiratory rate and often an increased thirst. Some acute cases will show classic diamond or raised red, firm swellings up to 2 inches across. In working boars the high temperature may kill off sperm reserves rendering him sterile or sub-fertile for six weeks after recovery.

Chronic:

The three long-term effects are:

• Skin sloughing of the 'diamonds' which is caused by an interruption of the blood supply. Ear tips and other extremities such as claws may drop off

• Osteoarthritis: Damage to the joints can be severe and permanent, leaving the pig irreversibly crippled.

• Endocarditis: The bacteria can circulate in the bloodstream where they can seed on the heart valves. They then grow producing the 'cauliflower' lesions, which eventually lead to heart failure, sometimes a year after the original infection.

Mild: In this form the disease often goes unnoticed as the pig may just go off its food for a day or be lame as a result of the joint inflammation.

This disease is so cheaply and so easily prevented with vaccination. Make sure the stock you buy are already treated with *Erysorb Plus,* for instance. If they are not vaccinated straight away, pregnant sows and gilts should be vaccinated three weeks before farrowing. This will give some immunity to the young piglets via the colostrum. Young stock must be injected with the Erysipelas vaccine at the correct timing, as stated in the manufacturer's instruction leaflet, then at three months old and of course, three weeks before the farrowing date. Thereafter as with all their stock every six months. Do not forget the boar!

Respiratory diseases

There are a number of viruses and bacteria that cause pneumonia, bronchitis and rhinitis (nasal cavity infection). Rapid or laboured breathing usually means a serious illness, so it must be treated immediately. Call your vet!

Bronchitis and pneumonia usually cause coughing in the young pig but may also be due to a lung worm infection. It is possible to vaccinate against all the above, Anthropic rhinitis with *A-RT* vaccine. I would emphasise though, that generally it is an unnecessary expense for the small pig keeper. It is best to keep your pigs in adequately ventilated, but not draughty buildings, and separate any infected animals immediately. It is a very good idea to get down to the pigs' level where they lie, to test for draughts; it is no good just leaning over the pig pen door.

Parvo-virus

This virus has been known for many years to cause reproduction failure, infertility and mummified foetuses in sows and infertility in boars, but these symptoms can also be caused by other factors. I would suggest, not vaccinating against Parvo-virus until a maternal blood test is submitted to a veterinary laboratory first. The vaccine is very expensive and the small pig keeper with just one or two breeding sows is highly unlikely to have parvo in his or her herd. Indeed even large commercial breeding herds are seldom routinely vaccinated against it, these days.

Foot and Mouth disease

Foot and Mouth disease is a notifiable disease and so must be reported immediately if it is suspected. It is caused by a virus and all cloven-hoofed animals are subject to it. Symptoms are loss of appetite, high temperature, lameness and some slavering at the mouth, although this is less noticeable in pigs than in cattle. There are blisters between the claws and sometimes on other areas of the body.

Most pigs recover from it and it is possible to vaccinate against it. At the time of writing, however, it is still official policy to slaughter all affected herds and to compensate the farmer for slaughtered animals.

Swine fever

Swine fever is another notifiable disease. Caused by a virus, it is most commonly found in young pigs. Symptoms are loss of appetite, thirst, shivering and possibly vomiting. A purple rash may be detectable on the ears.

All newly bought pigs should be kept in isolation for ten days and carefully observed in case they have been incubating the disease. The death rate is high. If it is detected, no movement of pigs to and from the site will be allowed. A preventative serum may be administered to the unaffected pigs to give them protection.

Mastitis

Inflammation of the mammary gland is sometimes seen in sows just before or after farrowing, especially if housed in damp and draughty buildings. Damage to the udder may occasionally be caused by the piglets' sharp teeth bringing chaps and cracks. If this happens rub in an antiseptic cream to stop infection.

If your sow develops mastitis the udder will be hot and red and obviously painful to her. Call the vet to get some penicillin into her quickly, with injections over three days of *Quartazone* to reduce the swelling.

Quartazone has another beneficial effect; it makes the sow thirsty, therefore she will continue to drink and make milk for the benefit of the piglets. It's a good idea to give the sow a hefty injection of vitamin B; this will stimulate her appetite so that she does not go off her food. Whatever you do, do *not* remove the piglets from her, as they will stimulate her milk supply.

I find that after two injections of penicillin it is best to alternate them with two injections of *Oxytetracycline* for a course of eight days. On occasion, the gilt or sow farrows with no milk. Injections of *Oxytocin* will certainly help but there is usually a good reason. Lack of milk or *Agalactia* is often the result of an unbalanced diet. Too much fibre and too little protein or little or no minerals and/or vitamins (usually A or D) so feed a good quality pig nut for the last four to six weeks of her pregnancy. Overweight sows tend to suffer from Agalactia. It is most important to make pregnant sows take enough exercise.

Farrowing fever

This cause of milk failure is thankfully not common now. It was a condition the 'old men' spoke of where pigs farrowed indoors in a hot, humid atmosphere. It does not happen when sows farrow outside in arks and are allowed plenty of exercise. There are characteristic symptoms, a whitish discharge from the vulva. She is reluctant to get up and her feet appear painful. She will have a temperature and her udder will be firm and hot. She will have *Endometritis* or inflammation, even *Septicaemia*, of the womb. A course of penicillin or streptomycin administered quickly should put her right.

Skin troubles

Bad skin, rough, reddened or flaky on pigs usually denotes bad management or incorrect pig feeding, or the presence of parasites. An old cure-all recipe is easily made up of one part oil of tar, two parts oil of linseed and then stir in flowers of sulphur to make a creamy ointment. Rub it on the affected parts, usually at the top of the front and back legs. (The recipe is also most-effective on dogs with eczema).

A universal and extremely useful skin conditioner is 'Pig oil'. One buys it from the local agricultural merchant. It is quite cheap and is a very sound investment as it can be used on pigs of all ages to very good effect.

Mange and lice

Pig lice are easily seen with the naked eye. They are like tiny crabs on the skin. They are often present in the ears. A pig will shake its head and perhaps hold its head on one side if they are present. The ears will have to be washed out with a mange wash or even just soapy water. It is essential to do this before your sow is put into her farrowing quarters. Pig lice eggs are yellow in colour and can be seen stuck to the ends of the pig's hair around the root of the tail, behind the ears and at the tops of the front and back legs.

Mange is a mite that burrows under the skin and is commonly seen in pigs of all ages if not treated. The parasite is the same sarcoptic mange that is seen in dogs and horses and nowadays more frequently in wild foxes. There is a reddening of the skin, which becomes rough and wrinkled and eventually develops into great crusts. The pig will become unthrifty and weak. It produces frenzied scratching in the pig and if spotted at this early stage must be treated immediately with mange wash.

If treatment is combined with washing, it is a simple matter to inject your pigs regularly or at any rate twice a year with the anthelmintic *Ivomec* or *Dectomex*. They both treat external as well as all the internal parasites. This treatment is highly effective.

Worm control

Worms in strong well fed pigs do not show up dramatically but in little pigs or newly weaned ones, those that show their back bones, are pot-bellied and have 'staring' coats and scour have worms. Sows that are excessively thin after or during rearing a large litter have worms, sometimes so badly it is impossible to get weight back on them. (Thin sow syndrome). The most serious effects are usually poor growth performance in growing pigs and condemned livers at the slaughterhouse because of the so-called 'milk spot'. Outdoor pigs face a greater risk of infection because worm eggs contaminate the pig paddock soil, sometimes for years. The vet will advise on the best course of treatment against worms. The *Ivermectin* wormers control all the categories of worms: stomach, lung, liver and intestinal. Ascard worms migrate through the lungs and liver sometimes causing pneumonia in young piglets. The eggs can last for many years in pig paddocks.

The regime is to treat all adult pigs, including boars, every 4 to 6 months. Treat sows at weaning, along with all the litter you've just weaned from her and then treat her three weeks before farrowing when you vaccinate against Erysipelas. Keep good records when treatment is administered and keep the treatment up.

Common disorders

The Trembles: This is sometimes seen in the fattening pig in the odd individual when it is eating. It is believed to be hereditary and no treatment is necessary. Do not keep the porker too long. Send it away to slaughter as soon as it is economic and certainly do not breed from it.

Ear Canker: Canker right down in the ear is due to grease and detritus which set up considerable irritation. More often the irritation is caused because of pig lice. If your pig continually shakes her head, treatment is essential. Tie cotton wool to a round-ended stick and swab the ear out with hot water and washing soda. When her ears are dry puff in some *Terramycin Aureomycin* powder.

Fighting: Where fighting occurs, a pig's bite is so toxic that the wounds frequently become infected. When this happens, bathe with salt and warm water. The belligerent old sow often attacks subordinates by biting their vulvas. This can be particularly nasty. Again bathe with salt water and puff on wound powder as with other wounds and separate the offending pig.

Lameness: The common cause of lameness in pigs used to be the aftermath of Erysipelas, before vaccination was widely used, with the result that rheumatism set in. Lameness does occur, of course from injury, and often in large sows, from sore feet if kept on rough ground or concrete for too long. Their soles can become so tender or raw that they will refuse to get up and they cannot be persuaded to walk. Their condition becomes progressively worse after each farrowing. The only course left to the pig keeper is to end the poor sow's pain by slaughter.

There is a form of foot rot in pigs, as in sheep. The diseased hoof should be cut away if possible and a foot rot spray or a copper sulphate ointment applied. Ample bedding is required for the pig indoors, or well-grassed paddocks that are changed before they get too cut up or frosted in winter. Avoiding bad conditions underfoot is the best way of preventing lameness.

Poisoning: It is very surprising that poisoning is so rare, considering that pigs have such a voracious appetite. They root in the ground and seem to eat anything they come across. I've never known any of my pigs to be poisoned. Over indulgence is much more common.

• Acorns are a good pig food but too many gives gastro-enteritis or cause abortion in pregnant sows. It is the shells that are dangerous because of their high tannic acid content. The antidote is to purge the pig with 1 - 2 fluid ounces of castor or linseed oil, depending on the pig's size.

• Bracken is generally held to be poisonous but pigs can be kept on bracken infested land with no ill effects, provided they have access to adequate water and are fed with their usual ration of pig nuts or meal.

• Foxglove poisoning is usually fatal. It affects the heart and causes excitement, quickened breathing and heartbeats, followed by a coma and death. I have foxgloves growing in some of the pig paddocks, however, but have never known the pigs to eat them.

• Rhododendrons are very common on acid soils and are said to contain poisonous substances but pigs seem to be highly resistant, especially adult ones.

• Hemlock can produce poisoning in pigs, either from the leaves or the root. It makes the pig froth at the mouth and stagger about. She may have fits and may die. Call the vet quickly in all cases of suspected poisoning.

• Fodder beet is an exceptional feed for pigs, being very high in energy making it an ideal feed during winter but it is *not* a good idea to feed the green tops because they can cause poisoning. This shows as a very watery scouring.

• Laburnum is very poisonous. It shows in pigs as nervous excitement then convulsions and coma followed by death. An antidote of tannic acid must be given extremely quickly if there is to be any hope of saving the pig.

• Yew is well known to be poisonous to all classes of stock. It is the leaf not the sticky berries that are so poisonous. The effect is usually sudden death and there is no antidote.

• Zinc poisoning is rare; this is only a word of warning. When milk or whey is stored in galvanised tanks there can be a reaction between the lactic acid and the zinc. If the milk or whey is fed to a lactating sow as part of her feed, the poison can pass to her piglets via her milk.

Showing

"Good conformation in the pigs selected for showing is essential, if you are to be anywhere near success."

I am a great believer in the showing of pigs, as there is no commercial yardstick like the show ring.

The success of pig showing starts way back with the selection and buying of the best stock of the breed you choose and in buying the best you can afford. I would suggest you go to a few agricultural shows, where there are pig classes and stand and watch the old hands show their pigs. When the showing is over, engage them in conversation. They are a friendly bunch, especially if you compliment them on their prize-winning pigs. Most, you will find, will be very helpful.

Most pig classes in agricultural show schedules are governed by age and, two months in particular; January and July. Pigs ought to be born, for instance, before the 1st July for a certain year, or after this date for another class. Again for instance, gilts must be born on or after 1st January. This is because a good bigger pig will always beat a good smaller one in the same class.

If one wants to do this showing job properly, you need to farrow your sows in January and July to obtain your 'Januarys' and 'Julys' as the showing people say. The next step is to pick out likely candidates and feed them well. They need enough protein in early life to lay down a good foundation and to maintain the base of good conformation.

Good conformation

Good conformation in the pigs selected for showing is essential, if you are to be anywhere near success. By good conformation, I mean good sound feet, well formed on strong upright pasterns. The legs should be straight and not turn out. The front ones should show no indication towards being knocked-kneed and the hind legs should not be 'sickle' or 'cow hocked'. The pig should move straight and freely on all four legs, with a complete absence of any lameness. A judge is within his rights to ask you to leave the ring with a limping pig.

Next comes the underline. Most judges start from the bottom up. The underline should be even, with the same number of teats on both sides, preferably fourteen. These should be evenly spaced and all well formed in a gilt. In a sow there should be no unsightly lumps.

A boar should have six teats before the sheath and eight after. It is essential they are all present and evenly spaced as it is the boar that passes to his offspring, any fault he may have regarding his underline.

The pigs you pick out for showing should have well filled hams, no sign of any

Gloucester Old Spots at the show. Exhibitors wear white coats and are equipped with sticks and boards for controlling the pig's movement. Straw bales have been used as perimeter fencing and some of the trailers used to transport the pigs are shown in the background. *Photo: Katie Thear*

Saddlebacks being exhibited at the Royal Show, 2000. Details of the class being shown are indicated on the notice, as well as the results from the previous class. The pig here is being controlled with a crook rather than a plain stick, while the board carries an advertisement for the sponsoring company. *Photo: Katie Thear*

sloping off from the hips. They should be rounded with the inner thighs well fleshed to the hocks when viewed from behind. The back must be level or even, slightly arched, which is a sign of strength. The body needs to be deep through from a line at the top of the shoulder to the chest behind the elbow and have well sprung ribs, with plenty of heart room.

The shoulders should be fine, not over-heavy and definitely not wider than the hindquarters. One likes to see a neat head with a minimum of jowl under the neck, with top and bottom teeth that meet. The skin should be smooth and free from roughness and sores, and with a silky coat.

Faults or points to avoid

Avoided should be any tendency for the pig to turn its feet out when walking. Splayed claws (especially in the front feet) and crooked legs or flat pasterns are equally to be avoided. Legs should not be too long and with a lack of hams, or slab-sided and narrow gutted. Any sagging of the pig's back is a definite fault, as are coarse shoulders and head. Over-full jowls and ears, with a tendency to be pricked in a lop-eared breed and of course, vice-versa, are aspects to watch out for. Under-shot or overshot jowls are to be avoided, and finally any swirl of hair on the back which we pig men call a *rosette*.

Show preparation

A month before the show, worm all the pigs you will be taking. I recommend the use of *Ivomec* or *Dectomax*. These clear up any exterior parasites as well as internal worms. This treatment will condition the pigs' skin prior to using *Pig Oil* a week or two before the show. Pig oil lifts old skin and detritus and renders the pig's skin and jacket smooth. If your pigs spend all their days outside, it is beneficial to shampoo them a couple of days before show day, to bring up the lustre of the coat. Shampoo them again the evening or morning of the show. Any shampoo or liquid soap will do. Some old hands use *Lux* soap flakes.

Before black pigs are put in the ring, oiling them up with pig oil gives them a lovely shine. White pigs are heaped with 'Woodflour', which is very fine sawdust. It is brushed off just before the pigs go into the ring.

I have assumed that you and your pigs have practised walking out and manoeuvring around the yard and through gateways with a pig board and pigstick or bat.

In the ring all competitors wear white coverall coats. Keep your eye on the judge so that you can show your pig off to its best advantage. It is necessary to be able to keep the pig still when the judge inspects her underline, which indicates how important it is to have practised at home.

Marketing

"Marketing is something to consider before buying the first pig".

This is a subject that needs a deal of thought before you put your gilt in-pig, or even before you buy a gilt. If you started with a couple of weaners before taking the next step in breeding your own, you might well have sold one of the weaners as pork to the neighbour, friends or relations. You will already have done the exercise of working out how much they cost to bring on to pork weight by the amount of feed they consumed. The abattoir's killing charge and what the butcher charged you to cut them up into joints, must also be added. You end up with quite a high figure. This will indicate the price to which you must mark them up.

All pigs raised for meat must be slaughtered at a licensed abattoir so they need to be transported in a trailer that is suitable for the purpose.

The novice pig keeper invariably makes the mistake of keeping his porkers too long, with the result they are either too heavy or too fat, or both! He then finds that customers are put off by the enormous quantity of meat and the inch of fat or more, on the chops. If you buy the quick maturing pork breeds as weaners, namely the Middle White or the Berkshire, then make use of the fact that they mature quickly, and do not keep them too long. They will not eat as much feed, and feeding them too long is a waste of money. The Middle White can be very economical; being a small pig it does not require as much food as a Gloucester Old Spots, for instance.

In my opinion the best way to sell pigs is as newly weaned weaners. I always recommend worming them at this time. Sell only healthy stock, not less than eight weeks old. You will find that you will quickly build a reputation for quality weaners. The 'old men' will tell you that*, "The first profit is the best profit"*, meaning they will have cost you very little, with little or no trouble and at a decent margin.

Do not be tempted to sell them in a market! There seems to be no trade for coloured *Slips* (weaners or stores) so sell them privately by advertising in the local papers. As your reputation grows you might find it beneficial to advertise nationally in a specialised magazine such as *Country Smallholding.* I certainly do. People come to me from all over the South to buy my Oxford Sandy and Blacks.

When you learn to get an eye for a good pig, buy pedigree stock, use a pedigree boar on them and sell pure-bred offspring. A pure-bred weaner for *breeding* can command twice the price of a weaner for *finishing*. It will have cost no more to rear to eight weeks of age, and with no more trouble. If they do not go at that age take them on to ten months old when you may sell. Unless you have your own boar, I would not recommend selling them in-pig. This would involve travelling, sometimes a good distance and leaving your gilt in the hands of someone you may not know, and whose boar may take sometime to get her in pig for various reasons.

If you keep white commercial type pigs you can try and sell your porkers to your local abattoir. They never pay a good price but will take your surplus at market prices. At least you will know where the pigs have gone, something you will not know for certain if you send them to market. Nine times out of ten they will end up in the same abattoir anyway with the added stress of going through the market.

Specialist butchers: With rare breed pigs there is the option of selling to a specialist butcher. Beware though, for most of them will not take the odd porker from you. If they do, they will often knock something off the agreed price because, *"It was much too fat"*. To really interest them you must guarantee continuity of supply and quality. Then, and only then, can you claim a premium price. This takes, as you can imagine, a certain amount of skill. It also brings you up a couple of steps, to twelve sows and a boar. Twelve sows farrowing twice a year, rearing 8 or 9 piglets a litter will produce 192 to 216 porkers a year, which will supply your butcher 3 or 4 porkers a week with a few more for Christmas. It also means a trip to the abattoir every week of the year, with the added expense of doing so. The abattoir will deliver the carcases to the butcher normally, and will charge him the cost of killing, offal disposal and levies, etc. Do not get lumbered with the total killing charge; going halves is the fairer way of going about it.

R.B.S.T. Meat Marketing Scheme: Another option is to sell your pure, rare breed porkers through the *Rare Breed Survival Trust Meat Marketing Scheme.* The *R.B.S.T.* also have a list of recognised finishing farms. These will take your weaners from you at a reasonable price but they must, of course, be pure-bred. The whole idea of the scheme is to encourage rare breed pig breeders to breed pure-bred offspring by giving them a ready outlet.

Farmers' Markets: The most satisfying and cost effective method is, of course, to market your own pigs. Selling them as meat, bacon and sausages brings the highest return because, apart from the slaughtering, cutting and bagging which must be done by a licensed slaughterer and butcher, there are no 'middle-men' in the equation. Farmers' Markets, where produce is sold direct to local customers, are becoming popular. They are normally held at specific venues and at regular times (eg, first and third Saturdays of the month). To sell in this way requires a good deal of organisation to ensure that there is no seasonal dip in supplies. It is necessary to be licensed where meat products are sold. See *Regulations*.

Organic production: There is undoubtedly a good market for organically produced meat, so pigs reared to organic standards will earn a premium on the value of their pork. Changing an existing herd to organic status requires a period of conversion, as the land on which they are kept needs to be organic and the herd needs to be fed and managed to full organic standards. There is now a common standard across the European Union. You can obtain full details of organic standards from

UKROFS or a licensing body such as *The Soil Association* or *Organic Farmers and Growers*. There are five key points regarding feeding:

• Feed should consist of 80% organic ingredients on a dry matter basis. You can mix your own feed, or there are suppliers of approved organic feeds.
• Fishmeal can be included in feed up to 12 weeks.
• No 'extracted' materials to be included in feed. These are things like soya which are extracted using solvents. Full fat soya is the accepted product.
• No genetically modified materials
• No medication, antibiotics, probiotics or medicinal levels of copper.

There are also organic standards covering housing and management. These are available from the certification body concerned.

Freedom Food Standards: *Freedom Food Ltd.* was set up by the *RSPCA* to ensure a high standard of animal welfare for farm livestock. Their standards for pigs can be obtained directly from them, and the meat verified to meet their criteria can be sold as meeting *Freedom Food* standards. It is not as exacting as producing meat to organic standards but does ensure that welfare considerations for non-intensive management are met.

Farm shop: Having your own farm shop is perhaps the ideal, but this is not feasible unless the scale of operations is sufficient to warrant the financial outlay. Local authority 'change of use' planning permission will normally be required, as well as building permission if a new building is to be erected. A licensed slaughterer and butcher will need to prepare and package the meat. The premises must also be registered with the *Environmental Health Department* where meat products are being sold so that inspection ensures that *Food Safety Regulations* are being met.

Organic meat being sold at Bath Farmers' Market. *Photo: Katie Thear*

Bacon

There are two methods of curing bacon described below. Sugar is used to soften the meat, and aromatic herbs and spices can be included as desired. You need cold temperatures so never attempt it during the summer.

Dry curing

Ingredients:
- 350g (0.75lb) sugar
- 570g (1.25lb) coarse sea salt
- 42g (1.5oz) saltpetre
- 2 crushed bay leaves

This will be sufficient for a piece of meat weighing 5.7 kilos (12.5lb), around ten times the weight of the salt. Saltpetre has been a source of concern in the past, but it is now approved by the organic certification bodies. It is not easy to buy, so there is a suppliers' reference at the end of this section.

Begin by rubbing the mixture into the meat, massaging it well in, then leave in a deep dish on a thin bed of the mixture in the refrigerator for 24 hours. After this, the blood will have run out and the meat can be laid on a fresh bed of the mixture and covered with the same. Leave it for around 3 weeks, turning it each day and rubbing some fresh mixture into it each time. Finally, hang it up to dry. Hang in a cool, dark, dry and airy place. If the area is not insect-proof, you can cover the meat with muslin, but only after it is completely dry. When it has hung for 2-3 days, the meat can be cooked, as desired. If not, it should keep for several months.

Wet curing

For a 5 kilo (10lb) leg of meat, you will need the following ingredients:

500 g (1lb) coarse salt.

For the brine: 750g (1.5lb) coarse salt in 3 litres (5 pints) of water
250g (8oz) molasses or soft brown sugar
1 teaspoonful saltpetre

Begin as with the dry cure. Rub the salt into the meat and leave in the refrigerator for 24 hours. Make the brine solution with the water, salt, sugar, saltpetre and some herbs and spices to taste. Rosemary, thyme, juniper berries, bay leaves and cloves can be used. Boil the brine mixture for 10 minutes, then leave to cool completely. Next day, brush the salt from the meat and place it in a non-corrosive container, eg, earthenware crock, stainless steel container, food-grade plastic bin, etc. Cover the meat with the brine mixture, put the lid on and leave in a cool place no warmer than 8°C (46°F). Check daily. If the brine mixture smells 'off' discard it and replace with fresh brine. After 2-3 weeks, remove the meat, rinse well and dry as above.

Smoking the meat preserves and adds flavour, but does not cook it. Whether you use a smokehouse or smoker, follow the instructions carefully to avoid carcinogens from the smoke. Wood chips used, for example, must be from hard woods.

Saltpetre can be obtained from: *The Natural Casing Co. Ltd, PO Box 133, Farnham, GU10 5EB. Tel: 01252 850454.* For smoking, contact: *Stamford Smokers. Tel: 01780 756563*

Reference section

Regulations

• **Registration**: Anyone with a pig (even a pet one) must notify the local *Animal Health Office* (AHO) of the local *Trading Standards Office*. (See local Telephone Directory)

• **Movement of Pigs**: The *Movement and Sale of Pigs Order* specifies that to transport a pig from one place to another, a movement licence must accompany the pig. This is called the *Schedule 1 Holding Movement Record.* Nowadays one can issue one's own licences. Blank copies are obtainable either from the AHO or by getting a book of tear-out licences from the nearest NFU office. A blank copy can be photocopied or you can type your own for producing copies.

SCHEDULE 1
HOLDING MOVEMENT RECORD
The Pigs (Records, Identification and Movement) Order 1995

Name and Address of person keeping the record . . . .
. . . .

Date of Movement	Identification Mark	Number of Pigs	Premises from which Moved	Premises to which Moved

No pigs can be moved off a site within 20 days of any pigs coming onto that site (not including the day of movement). Exceptions are pigs going to the abattoir, to shows, for breeding purposes or veterinary treatment. The AHO will provide further details.

• **Identification**: Pigs must be identified with an eartag or tattoo. If they are pure-bred, pedigree pigs, the breeder must have completed and submitted a *Birth Notification Form* to the *British Pig Association* by the time they are eight weeks of age. There must also be a number permanently marked on the ear. In the case of light pigs, this will be a tattoo, while black pigs and Tamworths have notches in the ear. Tattoos show consecutive numbers. Breeding pigs will also show the breeder's designated *Herd Letters*. Notching instruments are available from veterinary instrument suppliers and a little iodine can be applied to the ear when the the process is complete. Once it has been done, and there is a valid *Birth Notification Form*, it is then eligible to apply for registration in the *Herd Book* with the *RBST.*

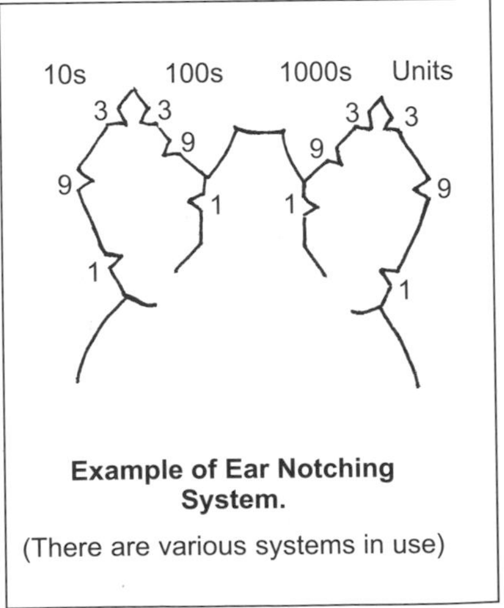

Example of Ear Notching System.

(There are various systems in use)

• **Transport of pigs**: The *Animal Health Act* and the *Welfare of Animals during Transport Order* specify the welfare conditions and requirements for transporting, loading and unloading of pigs. Trailers must be suitable for the number of animals and provide safe, non-slip access. The trailers must be cleaned and disinfected after use.

• **Slaughtering**: All pigs must be transported to an abattoir for slaughtering by a licensed slaughterman. The meat is inspected, butchered and packed according to the requirements of the *Fresh Meat Hygiene and Inspection Regulations.*

If a pig is a casualty of disease or accident, the treatment of the casualty pig is in the hands of the farmer. He may decide to despatch it to an abattoir or to have it killed on the farm. At the abattoir, the casualty must be accompanied by an *Owner's Written Declaration (Reg. 18(A) Schedule 18)*. Copies of this form may be obtained from your veterinary surgeon. It must be filled in by the owner with:

- Name, address and telephone number of the owner/person in charge.
- Name and address of owner's veterinary surgeon.
- Animal species, breed, age, sex and identification marks.
- Within the last 28 days has the above animal:
  - (a) received no treatment or
  - (b) received the following treatment:
    Give name of treatment drug, total amount and duration of treatment and the date of last treatment.
- Signs of injury, defect or illness the animal has shown or the vet's diagnosis.
- Signed by the owner or person in charge, with the date and time.

Unfortunately a pig may have to be killed on the farm using a suitable firearm. This is of course, provided the vet, licensed slaughterman, owner or stockman is familiar with the gun or pistol and holds a valid firearms certificate. The most effective and humane firearm is the 12-bore shotgun.

As a guide, the end of the barrel should be held 5-25cm (2-10 in) away from the head, on the diagonal lines between eyes and ears where those lines cross. The operative must have confidence that he can do it. If the pig's head is turned away shoot behind an ear pointing the gun towards the brain. An 0.22 rifle is not recommended as the calibre is too small (except for little pigs) and there is the risk of ricochet.

If you have any doubt in your confidence to dispatch a casualty pig humanely and swiftly you must call your veterinary surgeon to kill the pig humanely with an overdose of an anaesthetic or with a humane killer. Please take advice on all aspects of this rather distasteful subject. The vet or local *Animal Health Office* will be willing to help you.

Selling meat: Meat that has been prepared according to the *Fresh Meat Hygiene and Inspection Regulations* can be bagged and returned to the owner of the pigs for subsequent use or sale. To sell the meat, the producer must be registered with the local *Environmental Health Department* who will need to inspect the premises in order to ascertain that *Food Safety Act* requirments are met, as well as the labelling ones of the *Trades Description Act*.

• **Feeding of swill**: This is now banned.

• **Veterinary Medicines**: It is required to have a *Veterinary Medicine Administration Record Book* which can be inspected by *DEFRA* officers. An example is shown below:

Schedule 2 **Veterinary Medicine Administration Record**

The Animals, Meat and Meat Products (Examination for Residues and Maximum Residue Limits) Regulations 1991

Name and full address of person keeping the record ....

Date of purchase of veterinary medicine	Name of veterinary medicine & quantity purchased	Supplier of veterinary medicine	Identity of animal/group treated	Number treated	Date treatment finished	Date when withdrawal period ended	Total quantity of veterinary medicine used	Name of person who administered veterinary medicine

• **Notifiable diseases**: Some diseases, if suspected, must be notified to the authorities. They include Foot and Mouth disease and Swine Fever. A list is available from the *Animal Health Division A*, Hook Rise South, Tolworth, Surbiton, Surrey KT6 7NF.

• **Zoonoses**: Some diseases can be transmitted to humans and care should be exercised when handling pigs. (Hands should always be washed as soon as possible). Zoonotic diseases include Streptoccus suis, Brucellosis and Ringworm. The *Health and Safety Executive* has a booklet entitled *The Occupational Zoonoses* which provides further information. It is available from *DEFRA Publications.* (See below).

Insurance

Public Liability Insurance is recommended because in the event of an accident or damage caused by pigs, the owner is liable. It is also essential if the general public are coming onto the site, as for example, to a farm shop or a farm park. It is also possible to insure stock, buildings and equipment against accident, damage or loss.

Publications

Codes of recommendations for the welfare of livestock: Pigs.
Pig Identification, Records and Movement: A Guide to the Legal requirements.

Organisations

ADAS Woodthorne, Wergs Road, Wolverhampton, West Midlands WV6 8TQ Tel: 0845 7660085 - Government advisory service for farmers.
Animal Health Office - contact through your local *Trading Standards Department.*
DEFRA Department of Environment, Food and Rural Affairs (formerly MAFF) Nobel House, 17 Smith Square, London SW1P 3JR. Helpline: 08459 335577. Publications: 08459 556000
FAWC Farm Animal Welfare Council. 1A Page Street, London SW1P 4PQ Tel: 020 7904 6531
Freedom Food Ltd. The Manor House, Causeway, Horsham, West Sussex RH12 1HG Tel: 01403 223154 - *RSPCA* farm animal welfare scheme
Humane Slaughter Association. The Old School, Brewhouse Hill, Wheathampstead, Herts AL4 8AN Tel: 01582 831919 - Charity concerned with the welfare of livestock in markets, during transport and at slaughter.
MLC Meat and Livestock Commission. PO Box 44, Winterhill House, Snowden Drive, Milton Keynes, Bucks MK6 1AX Tel: 01908 677577

NFU National Farmers Union. Agriculture House, North Gate, Uppingham, Rutland LE15 9PL Tel: 01572 824686

National Pig Association. PO Box 29072, London WC2H 8QS Tel: 020 7331 7650 - Organisation representing the pig industry

Oganic Farmers and Growers Ltd. 50 High Street, Soham, Ely, Cambridgeshire CB7 5HF Tel: 01353 722398 - Organic certification body.

Pig Veterinary Society. Southview, East Tytherton, Chippenham, Wilts SN15 4NX

Rare Breeds Survival Trust. NAC, Stoneleigh Park, Warwicks CV8 2LG Tel: 024 7669 6551 Conservation body for: Berkshire, British Lop, British Saddleback, Gloucester Old Spots, Large Black, Middle White and Tamworth. Meat marketing scheme for rare breeds through accredited butchers.

The British Pig Association. Scotsbridge House, Scots Hill, Rickmansworth, Herts WD3 3BB Tel: 01923 695295 Registration body for the following pure breeds: Berkshire, British Hampshire, British Saddleback, Duroc, Gloucester Old Spots, Large Black, Middle White and Tamworth. (See also Breed Societies)

The Soil Association. Bristol House, 40-56 Victoria Street, Bristol BS1 6BY Tel: 0117 914 2400 Organic certification body.

UKROFS United Kingdom Register of Organic Farm Standards. Room 320C c/o DEFRA, Nobel House, 17 Smith Square, London SW1P 3JR Tel: 020 7238 5915

Breed Organisations

Berkshire Pig Breeders' Club. Mrs Barnfield, Blaisdon House, Aston Road, Kilcot, Newent, Glos GL18 1NP Tel: 01989 720584

British Kunekune Pig Society. Banns Cottage, Nomansland, St. Buryan, Penzance, Cornwall TR19 6EL Tel: 01736 810519

British Lop Pig Society. Miss M.J.Hadley, c/o Chesterton Fields Farm, Fosse Way, Leamington Spa, CV33 9JY Tel: 01926 651158/661547

Iron Age Pigs. D & E Sebborn, Green Acres, River Lane, Dunwear, Nr. Bridgewater, Somerset TA7 0AA Tel: 01278 451864

Large Black Pig Breeders' Club. Mrs Sue Barker, West Farm, Ruckley, Shropshire SY5 7HR Tel: 01694 731318

Middle White Pig Breeders' Club. M. Squire, Benson Lodge, 50 Old Slade Lane, Iver, Bucks. SL0 9DR.

Oxford Sandy and Black Pig Society. Tadneys Farm, Fox Lane, Kemsey, Worcs WR5 3QD

Tamworth Breeders' Society. Caroline Weatley-Hubbert, Boyton Farms Co. (Wilts) Ltd, Boyton, Warminster, Wilts. BA12 0SS.

The British Saddleback Breeders' Club. Carole Muddiman Tel: 01280 850677

The Gloucestershire Old Spots Pig Breeders' Club. Richard Lutwyche, Dryft Cottage, South Cerney, Cirencester, Glos GL7 5UB Tel: 01285 860229

Suppliers

Feeds

Allen and Page Tel: 01362 822902

W & H Marriage & Sons Ltd. Tel: 01245 354455

Outdoor Pig Housing

Carbery Plastics (Ireland) Tel: 00353 23 33531

Farmwright Tel: 01805 623255

John Booth Engineering Ltd Tel: 01903 716960

Ramthorne Farm Pig Arks Tel: 01608 730640

General Suppliers of Pig Equipment
Atlantic Superstores Ltd. Tel: 0845 4584745
Ascott Smallholding Supplies Tel: 0870 443 0653
Clarkes of Walsham Tel: 01359 259259
Glendale Engineering Tel: 01668 216464

Courses: are available from some Agricultural Colleges, *Small Farm Training Groups* or *Smallholding Associations*. There are also some farms that offer hands on training for beginners, viz: *Piggy Weekends* (Devon) Tel: 01404 831310

Breeders: Lists of breeders can be obtained from the *RBST* and the *British Pig Association. The Breeders' Directory* in *Country Smallholding* magazine is also very useful. (See also their website at www.countrysmallholding.com

Rules of thumb

• Pig people are quietly spoken, kindly folk who work exceptionally hard.
• Buy the best pigs you can afford.
• Always take the sow to the boar for mating.
• The small one that didn't make pork weight when his mates did can happily be mixed with a batch of weaners. He won't hurt them, as they are no threat.
• Clean up any food that weaners leave and give it to the older pigs.
• Learn to feed by eye.
• The keys to good management are observation and attention to detail.
• Learn to know when a pig is ailing before it knows it is ill itself.
• A pig's temperature is 39°C (102°F).
• A pig's heart pulse is 60 to 80 per minute.
• A pig's respiration rate is 20 to 30 per minute.
• A sow cycles every three weeks.
• A sow's gestation (from conception to farrowing) is 3 months 3 weeks 3 days (or 108 to 120 days).
• A sow comes 'hogging' to the boar from 4 to 8 days after her piglets have been weaned from her.
• A pork pig, when sent to the abattoir has a killing out percentage of about 72%.
• Pig feed amounts to 80% of the pig farmer's costs.
• It takes about 306 kilos (6 cwt) of meal to finish a pork pig.
• It takes about 204 kilos (4 cwt) of pig nuts to feed the sow from conception to farrowing; about 115 kilos (2.25 cwt) for little pigs like the New Zealand Kunekune.
• Keep it simple; enough can go wrong without making it complicated.
• Learn the pig's language: the meaning of all their grunts and squeals.
• When fostering piglets, put only the biggest and strongest ones onto the foster sow.
• The 'old men' will tell you that pigs with the thickest root to their tails make the best breeding stock.
• They also say, *"Dogs look up to you, cats look down to you but pigs are equal"*.
• You are never going to make a fortune.
• Remember that you are are in it because you love pigs.

Glossary

AI	Artificial insemination
Ad lib feeding	Allowing pigs free access to feed
Ark	Outdoor, moveable house
Bacon pig	Larger weight pig approx 90kg (198lb) live weight
Back fat	Depth of fat along the back
Bloom	Visual sign of good health
Boar	Intact male pig
Carcase	Dressed body of pig
Colostrum	First milk after farrowing
Conversion	Ability of a pig to convert food into a carcase
Creep	Piglet nest
Cross breeding	Mating of two different breeds
Deadweight	Weight of dressed carcase after slaughter
Farrowing	Giving birth
Finish	Last stage of fattening
Fold unit	Small house with fenced area for controlled grazing
Gestation	Period of pregnancy, between 110 and 116 days
Gilt	Female pig before first farrowing
In-breeding	Mating closely related individuals
Lactation	Period when female produces milk
Line breeding	Repeat breeding to a single individual
Litter	Offspring of single farrowing
Liveweight	Total weight of pig
Pedigree	Registered pure breed
Pork pig	Small size pig approx 62.5kg (140lb) live weight
Scouring	Diarrhoea
Sow	Female pig after farrowing
Store pigs	Pigs kept for breeding or future fattening
Swill	Waste food fed to pigs - now forbidden
Swine	Pig - old term
Vulva	External part of female genitals

To estimate the dead weight of a pig

Provided the pig stays still, measure the length in inches; that's from top of his head, between his ears, to the root of his tail. Multiply by the girth immediately behind the shoulders. Now divide the results by 13, 12 or 11 depending on the pig's condition (whether lean, medium of fat). The result is the dead weight in pounds. (1in = 2.5cm)

Temperatures for pigs

Pig	*Temperature* ^{0}C
Sows	15-20
Sucking pigs in creeps	25-30
Early weaned pigs	27-32
Weaned pigs (6 weeks & over)	21-24
Finishing pigs (porkers)	15-21 (baconers 13-18) (heavy hogs 10-15)

Index